AF693498

REVIEWER COMMENTS

From a free self-thinker, who has read this book

Robert Q. Riley - Did not graduate high school. Instead, he wrote articles on the contemporary design of cars and `boats for "Popular-Mechanics" magazine for 30 years. His last car design did 225 miles per gallon, in a world race of the most economical car of all cars.

ROBERT Q. RILEY was born in an Irish Catholic religious family; the result is free self-thinking and debating in scientific reality!

ROBERT Q. RILEY became a genius.

I've read this book on wisdom and do not see any contradictions in it. However, people perceive it as a contradiction that could come from their refusal to give up traditional religious ideas and beliefs.

Energy gives life and with it moral and humane values. We do not have to create God in our image for God to exist. All of us and the entire universe itself consist of God-energy. God is not an extremely powerful man or a woman. God is the life force of the universe, which creates life energy, in the cosmos as a platform for life.

Also, there is no difference between the "life-force" of the universe and the "energy" of the universe. If that were not the case, it would be impossible for these rocks (this stuff we think of as raw matter) to come alive and begin to think and create on their own as individual human-beings and as societies. Consider all of it as connected and consisting of the same energy. Free self-think and be creative for a better world in scientific debate conversations.

ROBERT Q. RILEY

THE ENEMY OF SCIENCE IS NOT IGNORANCE

It is Blind-faith Addiction

THE ENEMY OF SCIENCE IS NOT IGNORANCE

It is Blind-faith Addiction

LEON D. HALFON

Halfon Book Publishing

Los Angeles, California

THE ENEMY OF SCIENCE IS NOT IGNORANCE

It is Blind-faith Addiction

Published by Halfon Books
Los Angeles, California
Leon D. Halfon, Publisher/Editorial Director

ISBN: 979-83-057121-2-4
Library of Congress Control Number: 2025900130

DARKNESS IS 0, OR NOTHING.
THERE IS ONLY LIGHT.
DARKNESS IS A PLACE WHERE THERE IS NO LIGHT!

Certain long-held belief traditions are created by preaching lies and propaganda scams by atheist criminal leaders to create blind-faith gambling addiction, for people who want to believe that they are winners; and, as with any gambling addiction, they are accepted by the medical field as a disease, which are all hard to cure; and that makes you a loser, not a winner! However, believing in blind-faith gambling addiction in the darkness of memorizing knowledge, rather than turning on the light to see the creator. God, by any other name, is still the creator, in the light of scientific reality that is there; it seems like frightening challenges, but it is essential for change. You might feel more comfortable in the blind-faith darkness. But once you switch on the light to reveal the universal truth of the time, you will see there is nothing to be afraid of. At that moment, you will accurately understand that it is more frightening to live in a dark, blind-faith illusion of religion than in the light of the creator in the reality of science. You will be excited to realize you can free self-think to understand things clearly in reality to realize that life is not a monotonous, boring journey of comfort, by competing illusionary beliefs, opinions, philosophies, and fairytales, created by cruel selfish (atheist) Leaders for more money, by making you an idol worshipper, unsafe and unhealthy! Because God is only found in your infinite mind, not in your computerized brain. Leaders want you to be "moldable" because they are greedy from competing egos by memorizing knowledge. Therefore, the government is the entity that needs to be moldable, if it is to represent the creative people, which will direct this country to direct the rest of the world—in the right scientific direction of reality. **Once you understand that there is a Creator of the Energy, no leader, cruelty, or propaganda is capable of making you not have confidence or demoralize or make you immoral. However, the word belief creates no confidence, no certainty, no understanding, and that creates non-believer atheists, criminals, leaders, and blind-faith gambling addiction; and that is the reason society needs a lawyer and a big social government for every 100 people.**

So please don't tell me that religion makes people moral.

Challenge Yourself and your children to Excellence,
Improve Your Minds Like the Greatest
Minds by Creating a More Accurate
Self-Understanding of Life Reality of science,

Once you and your children understand that there is a creator, no leader, cruelty, or propaganda is capable of making all of you to not have confidence in yourself or to demoralize or make any one of you, immoral.

To create a better government, a better country, and a safer world, not by philosophies or illusions of knowledge, to keep grownups childish, to participate in hate, and war, from believing religious fairy tales, to feel important... and that is childishly stupid!

If you want to believe, you can believe anything, or anybody.

The point is to understand the scientific spectrum reality, by free self-thinking and accepting the possibility of being wrong in a realistic debate to better yourself and the world.

In this world, there is only change between two powers.

1. The original Babylonian Atheist leaders and royalists, that didn't want to lose their privileges and commoners, ignorant beliers obedient followers in blind-faith addiction darkness of memorizing knowledge, and that is hard to cure!
2. The new democratic movement, for all the people with a representative of the people in government, to free self- think and understand that there is a creator in the reality of science to be creative for a utopian creative world

Leon D. Halfon

This edition is created from the original text, and by debating and researching in reality of science, it developed into a more accurate understanding, for the wisdom of this new addition. However, knowledge, religion, and philosophy stay still, and universal-truth wisdom changes with time, to create new knowledge. Because life is meant to be continuously exciting and challenging, to have the understanding for the confidence to better yourself, your government, your country, and your world, to be safer and healthier.

DEDICATION

When I arrived in America 60 years ago from Israel, I could not remain close to most of my Israelite friends because I was a free self-thinker Hebrew searching **for universal-truth reality of that time,** to create the wisdom to make a better world. My Israelite friends were afraid and unwilling to step away from the dangerous discriminatory dogma of world "gangs" that created illusory religious philosophical beliefs by Babylonian magical religious leaders and to be united with the free self-thinkers of the world. So, as a free self-thinker, I became much closer to my non-Israelite new friends from quite different countries and backgrounds, to unite the free self-thinking of the world. These free self-thinkers were all in the same scientific way of searching for the wisdom to arrive closer to a universal truth of that time. This journey brought us all together and, from accurate understanding, we became united joyfully and trustworthy as my best friends. They are the ones to whom I promised I would write this book. But unfortunately, they are all gone from my life now, in one way or another and I miss them very much.

I wish to dedicate this book to my friends, who were humane-beings. These were the best free self-thinkers and scientific debaters in life reality searching for universal-truth to understand the magic of God's reality of science, more than anyone could have ever known: Maurice Braslave, Robert Bardey, Gerard Alcon, Jon Michel, Monsignor John Sheridan, Suliman and Jack Brady.

This scientific system always worked, and it will unite all the people on this planet and create a creative peaceful humane world in a democracy, to save this planet from destructive people that memorize knowledge and become cruel leaders and Royalists!

TABLE OF CONTENTS

Dedication xiii

Preface: What This Book is About 1

Introduction 2

PART ONE: WHAT THIS BOOK IS ABOUT 11

Chapter 1: Free Self-Thinking 12
Two scientific Experiments 16

Chapter 2: Traditions 17

Chapter 3: Wisdom 19
Religious Jokes looking in the Wrong Places 20

Chapter 4: The Socratic Method 22

Chapter 5: Masons from the Crusaders 25

Chapter 6: United States of America: The Greatest Country in the World 27

Chapter 7: The Most Creative Country in the World Used to be the United States of America 29

Chapter 8: Discover Free Self-Thinking in Search of universal-truTh 32

Chapter 9: Experiment to Know the Percentage of People Who are Humane-Beings 37

Chapter 10: What We Know Now from Higher Consciousness 39

Chapter 11: Example of Free Self-Thinking in a Debate 42

Chapter 12: Cognitive Dissonance 45
Understanding the Meaning of Words 46
Confidence 47

Illusions ... *48*

Chapter 13: Free Self-thinkers Can Recognize Right From Wrong ... 50
Obstacles of the Brain ... *51*

Chapter 14: Voodoo Zombies ... 54

PART TWO: YOU ARE WHAT YOU DO, NOT WHAT YOU BELIEVE YOU ARE! ... 57

Chapter 15: A Statement of Wisdom ... 61

Chapter 16: Hippies ... 63

Chapter 17: Satan, Lucifer, and the Devil ... 65

Chapter 18: Comparing Cultural Contributions of Religious Followers and Free Self-Thinkers ... 67

Chapter 19: Public Coffee Shop of Wisdom ... 69

Chapter 20: People of the Books ... 71
Ancient Books of Wisdom ... *74*

Chapter 21: Competition ... 75

Chapter 22: To Be a Humane-Being or Not to Be a Humane-Being - That is the Question? ... 77
Choose Conversation in reality instead of Violence ... *78*

Chapter 23: DAAA ... 79
Verifiable Conclusions ... *81*

Chapter 24: The understanding Movement ... 83

Chapter 25: The only Bible Story about Satan is in the Book of Job ... 84

Chapter 26: Discover if Someone is a Free Self-Thinker searching for universal truth ... 86
People With Little or No Formal Education ... *86*

Chapter 27: Discover A New Meaning and A New Word ... 90
Police Armor Shield ... *92*

PART THREE: EVOLUTION 93
Introduction: Science and Nature 94

Chapter 28: Free Self-Thinkers are Problem-Solvers 96
Free Self-Thinkers Need Artificial Intelligence..................................98

Chapter 29: What is Love? 102

Chapter 30: Is life a Glass-Half-Full or Half-Empty?..............................105

Chapter 31: Why There bullies? 107
Declare Your Brain and Mind Independence..................................109

Chapter 32: My Verifiable Conclusion Regarding Moses.......................111
More of Moses' Inspiring Wisdom Stories.......................................116

Chapter 33: My Understanding, from Debating the Story of Jerico........118

Chapter 34: King david..119

Chapter 35: Wisdom of King Solomon 120

Chapter 36: Abraham..124

Chapter 37: Hierarchy and Joan of Arc 126

PART FOUR: GOING BACK HOME 129

Chapter 38: Wisdom Stories about the Island of Djerba 130

PART FIVE: INSPIRING PHOTOS AND LIFE STORIES OF LEON.. 143

Chapter 39: Leon's famous Life Stories..144

Chapter 40: Leon's Inspiring Photos 150

Chapter 41: Leon's Favorite Animal Stories 170
A Young Appaloosa Stallion at My Ranch..171
My Thoroughbred Stallion 171

Chapter 42: Leon's California Stories And More...................................173
Swinging onto the Balcony..173

I Date a Woman Who Likes to Drink at a Bar ... *174*
Working in the Litton Building for Maurice Braslave ... *175*
The Voice of Europe ... *177*
The Preachers ... *177*
The Mustang Ranches ... *177*
My Cancer Stories ... *178*

Chapter 43: Living on The Desert Island of Djerba ... 180
Crushed ... *185*

Chapter 44: Leon The Lifeguard ... 193
Yoynay ... *198*
England ... *200*
America ... *201*

Chapter 45: The Second Gun ... 210

Chapter 46: FHA Foster Home Program ... 212

Chapter 47: Desilu ... 217

Chapter 48: Leon's Many Friendships ... 219
The Catholic Church ... *219*
Jack Brady ... *224*
Buzz Aldrin ... *227*

Chapter 49: Halfon-Dula Family History ... 228
My Brother, Yehuda ... *234*
My Mother, Rachel ... *235*

PART SIX: LEON'S FREE SELF-THINKING IDEAS ... 237

Chapter 50: Inventions ... 238
Solar Water Distillation System from Humidity ... *238*
The Best Sprinkler System ... *239*
Remote Control Drone ... *239*
Inexpensive Solar Water Heater ... *241*
Producing Electricity with Nature ... *242*
Ethanol at Five Cents a Gallon ... *242*
Build A Concrete HollOW Wall ... *242*

Converting to a Non-Skid Bathtub *243*
Replenish the Dead Sea in Israel *244*
Solar Air-Conditioning *244*
The Solar Energizer Necklace *245*
A Humane Non-Fatal Mousetrap *246*

Chapter 51: Free Self-Thinking creates Words of Wisdom by Leon 247
Lifeguard Poem *247*
A Free Self-Thinking Prayer for Life to care *248*
The Word Belief *249*
The Word Opinion *252*
Religion Is a Placebo *253*
Religious Leaders are Atheists *256*
The representative of the creative countries of the world is playing a game of chess for world domination *260*

Chapter 52: Betrayal and Other Wrongdoings 262

Contact Information 265

Notes 266

PREFACE

WHAT THIS BOOK IS ABOUT

This book is about the search for new universal truth in science, to better us and the world. When you know how the magic reality of science works by free self-thinking, you will understand how to love everything and everybody, and to live a joyful long life with a smile!

When you memorize the illusion of knowledge and become an obedient believer of a leader in blind-faith, you will blame everything on everybody, in hate and war, to live a miserable short life!

Any fool can memorize knowledge or religions, or philosophies, or preachers' propaganda from millions of professors that become leaders from the need to be important. The point is to free self-think, and by debating in scientific reality to arrive at a more accurate understanding and save this living planet from obedient followers in blind-faith addiction that is created by cruel selfish leaders who lie. From modified illusionary stories of the past, which keep grownups as idols and childishly stupid, and as extreme, scary inhumane torturers. Atheist leaders protect their basket of money because their god is money. (They are god players.) It is just a belief, not reality of science.

This book welcomes disagreement, to learn from each other in scientific reality, in search of new universal truth for the time, to find newer universal truth knowledge, to better the world for all! (323-449-2325).

This book does not tell you what to think, it will make you understand in your mind beyond any doubt, that there is a Creator of the Energy. God, by any other name, is still the creator.

The enemy of the Atheist people's parasite leaders 'that lies', are the people who search for universal truth, and that is science.

INTRODUCTION

In the past all food was registered as food to eat; however, some is poison, some is organic, and some is a vitamin. Therefore, today knowledgeable people don't accept the worst food to eat as good enough and need an accurate understanding of all the ingredients they put in their stomachs. **This is why they live longer!**

In the past, knowledge was considered and registered as a philosophy or a belief; however, today, people searching for universal truth in a scientific debate of reality, will eventually find the answer, **understand how to love all lives, live longer, and prosper.** However, if you are searching for truth, in religion or tradition in blind-faith addiction or philosophy or propaganda, in the illusion of knowledge that will keep your idol, and make you an idol worshiper, you will never find the universal truth, because it is subjective to each other or each group; and everything subjective to each other, or each group is impossible to be the truth, and believing in lies is poisonous in your brain. That is why people must, free self-think in the reality of science and debate, for a healthy brain.

If universal truth does not exist, then science cannot exist. For example, in physics, for every action there is an accurate opposite reaction and that is a universal truth!

How would you make this world better for life? If you were the Universe Energy, the creator God?

The universe is Energy, and the earth is heated from the outside by the Energy of the Sun, and from the inside by the Energy of the earth, from a scientifically accurate distance, for life to flourish. This is the reality of science, whether you believe it, or know it, or not.

Everything evolved from—or was created by—Energy!

Would you make the planet a sphere and give it gravity so that life would not fall off the edge? Would you make water boil at 100° Celsius, so water vapor could evaporate in the atmosphere to create rain so that life could exist on land and under the water? Would you make water float when it freezes at 0° Celsius, creating large ice floats that insulate the waters below so that life can thrive under the ice and on top of the ice floats? Would you make the ice white to reflect the sun's rays to not contribute to global warming? Would you make glaciers, snow melt, and water evaporate to create rain, and lightning to create negative-ions that clean, and refresh the dirty air, to become delicious, healthy air for life? Would you make the rainwater clean the dirty ground, to come out as delicious healthy spring water for life, and that creates rivers, so that life could have healthy water on land all year long? Would you have volcanoes thrust out all the minerals that enrich the earth for agriculture and the vital materials we use like copper, iron, gold, diamonds, and other elements to the surface, so life and people could utilize them and evolve? Would you have stars to explode to create carbon, which is vital for life on Earth? Would you make a magnetic field and ozone to protect life from the blasts of solar winds? Would you have earthquakes shake people from their comfort zone, and from illusion of knowledge to free self-think in God's reality of science and be creative? Would you create a planet just the right size to have perfect gravity so that people can walk upright on two feet? Would you make a planet with a tilted orbit to create four seasons in a year? Would you make a moon to light up the sky at night and keep the oceans' high tides and low tides flowing and the seas in constant motion, , which also influences our weather, **because life was ment to be challenging, inspiring, and exciting.**

Life needs a Goldilocks' atmosphere to flourish, not too hot, not too cold, but exactly right—with the right mix of gases, to breathe and thrive. Today you are able to see from the telescope, in the Milky Way, energy creating elements to create a planet for life. Would you create a super-heated planet named Mars for us to observe and learn from what happened to it due to global warming? Mars was a very earthlike planet long before an expanding Sun released excessive carbon dioxide into the atmosphere, creating a greenhouse effect that heated the surface temperature to over seven hundred degrees, hot enough to melt lead! Will that be Earth's fate someday if we do not become caring humane-beings to create wisdom? Would you make time so that life would be in constant motion? Would you give people only so many years to live and know that they are running out of time to accomplish something worthwhile by challenging their life to create a more accurate understanding?!!!

The father of psychoanalysis, Dr Freud said, "The most important thing for people is the need to feel important, and that is the second Universal truth in science that does not change whiz the time." It is the reason people memorize illusions of knowledge and knowledge to become important! That creates Atheists and leaders. However, all animals are Athiest, only by free self-thinking and debate in the search of universal truth in reality of science, you will be part of everything, and that will feel and be more important, for a continuously better life. **So, we all can be important and creative in science, to create a better world!**

Newton was a physicist and a mathematician, who changed the world for the better. He knew that the actions made by the universe's energy caused the reactions that created the universe and made life possible. **He called that action God. God is a word used to name the Creator the Energy.** Newton did not know how God worked. Newton wrote about his discovery but had to hide it, because he was afraid of being persecuted by inhumane cruel criminal religious leaders and their ignorant mobs of followers and believers. It took two hundred years after Newton's death for someone to find his book.

Here is an example of how free self-thinking correlated with a universal truth: Four thousand years ago, Moses, the people's representative, said that God is Energy. God does not make mistakes; people make mistakes. This is how Moses, the free self-thinker, described the wisdom and Energy of the Universe. God, "the Creator the Energy," is not a religion or tradition; it is just a word to name the cosmos' life energy. Everything around you and in you correlates with today's scientific knowledge. Look at the tree. It has no pump and no eyes. **However, it can find and carry water up from the roots below to the very top of the tree;** this happens through the power of life's energy. This is a reality not a belief. When I confront a challenging task, I call on the life Energy to correlate and accomplish the job.

In a television show about Space-Time on PBS's Nova TV program, Stephen Hawking said that God's energy of the cosmos is *in your subconscious*. Hawking demonstrated that the subconscious brain *knows* what you will do before the conscious mind acts in a test. What Stephen Hawking meant was that there is no free will because your subconscious decides before you do. As a free self-thinker, I do not agree with Stephen Hawking that there is no free will, because he miscalculated people's different speeds.

Speed 1: Instinct operates at the speed of light.

Speed 2: The subconscious moves at the rate of instinct.

Speed 3: The conscious mind is slower than the subconscious instinct because the mind must understand the brain before acting, and by free self-thinking or debating, the accurate answer could take from minutes to years by your free will.

Atheist people repeatedly said to me that after seeing all the terrible things that happened to people by leaders of the world, and God did not interfere, that's the reason they became atheists.

My response is, the reason God does not interfere, is to give you free will! So, stop memorizing knowledge and start free self-thinking in the search of universal-truth, in reality of science, and create a majority of the world, so we can stop these horrible god-players. **It is up to your free will.**

In 2012, scientists confirmed the long-sought Higgs Boson particle's detection, commonly known as the (God particle,) at the Large Hadron Collider (LHC), the most powerful particle accelerator on earth**, at the length of 4 miles. The Higgs** (God particle) **mechanism gives mass to fundamental particles, such as electrons, protons, and other building blocks that cannot be broken into smaller parts. However, these Higgs mechanism vital energy particles comprise only a tiny portion of the total mass in the Universe.** According to scientist Peter Higgs, "in the beginning, the universe was in perfect symmetry." Then, along comes the Higgs God particles' energy field that creates a super-symmetry, and that facilitates continuous change. Just like Earth, the Universe evolves, ever-changing, ever-expanding, universal energy powers.

Our planet's natural law for wildlife needs to evolve in diverse ways via continuous change to enjoy and be amazed by the magic of life. Would you create **love** to create beautiful species, which often must compete to procreate, sometimes violently, leading to a better and more beautiful species? **The crueler you are, the more you are an Atheist animal. The more you are humane, the more you become a higher consciousness.**

The scientist Nel Bohr discovered that measurement changes everything. He said that before a particle is discovered and measured, its characteristics are uncertain. For example, a particle could potentially exist in multiple places around a nucleus at any given moment; however, the act of higher consciousness measurement is what forces a particle to relinquish all other possibilities to one definite location. During the act of observation—and only at that moment, a particle's location becomes certain. Hence, this universe is designed for higher consciousness. From consciousness comes awareness. From accurate awareness comes brilliance.

People and animals have common sense, and logic, and are able to memorize knowledge to become intelligent, but only humane-beings of higher consciousness can become brilliant because they exercise their minds in the reality of science, by free self-thinking and debating. When a highly conscious mind observes, the magic of life collapses everything; however, when you look at any magic, you have to look at the magic by debating to see the spectrum for an accurate understanding. Since we can acknowledge ourselves and others around us, with a more precise understanding of creating a higher level of consciousness than another being on the planet, it is essential to free oneself by not believing in illusion of knowledge (religion) or Atheist philosophy, to evolve into higher consciousness by searching for universal truth to understand the Creator's Energy of life.

What is the difference between people of higher consciousness and being? We can send people to the moon. A dolphin, on the other hand, cannot send another dolphin onto land. So, what is it that makes us so brilliantly self-aware? Why are we capable of understanding things like God's universal truths of science and wisdom and why are we capable of having this "higher consciousness?" Because we are here on earth for an inspiring purpose and brilliant enough to realize in the spectrum of the magic of life that creates the wisdom to know **that we are responsible for protecting the environment and all life. Not because it is easy, but because it is a hard, challenging, inspiring, and exciting.**

Some knowledgeable people with no wisdom theorize that consciousness is an illusion created by chemicals in the brain. They can see where thoughts occur in the brain using Magnetic Resonance Image scans (MRI) and can even estimate what you think. Still, they cannot tell where you are or where your mind is; although they have figured out where activity in the brain operates, sculptors do not know who or what is in charge. Essentially, they cannot find "you or your mind" in the brain! On the other hand, as free self-thinkers in wisdom, we realize that there are things that people are not supposed to, or able to measure or know. However, only by an accurate understanding of the spectrum of the magic of life, you create understanding!

Be a free self-thinker grownup in reality, for a safer humane world! Because scholars are able to memorize the knowledge, they will compete to create a parasite hierarchy; therefore, they are not humane-beings, they are just beings, that is the reason they don't accurately understand the magic of life, because **people who memorize knowledge and compete to become parasite hierarchy, that will not comprehend infinity in their own Brain**, because the brain is an organ and all organs die. However, a free self-thinker in the search of God's universal truth of science, understands that universal truth continually changes with time to create an infinitely exciting and challenging life, and that understanding in your mind creates a higher consciousness humane-being, to understand that you are infinite. Because the energy of the universe will not have it any other way, the reality of the universe we live in has energy throughout, which provides all the necessary resources **to fix itself or destroy it!**

A higher conscious humane-being is a responsibility, not a privilege, and by his accurate understanding, has confidence in what is right versus wrong. Along with this wisdom comes energy and the willingness to accomplish the impossible. You can overcome pain, control your feelings, and conquer the most incredible obstacles in life by free will from free self-thinking **in your mind.** I have known this since I was a little kid. Once I understood the mechanics of higher consciousness from free-self-thinking wisdom and the energy it offers, I could harness life itself and become magically powerful! I feel everybody should learn the powers that exist inside of them and around them. They too, can have the energy and confidence to love themselves and all lives, to accomplish wonderful things. Unfortunately, not everyone is open-minded to free self-think for more accurate thinking of wisdom because of obstacles in the brain from memorizing the illusion of knowledge, and knowledge,

This establishes in you an understanding mind beyond a shadow of a doubt, that there is a creator. God by any other name, is still the creator. That is a good reason to be humane and not believe in illusions of knowledge, because all this could not have happened by superstitions or delusions of knowledge or ads or %, or by oxidant, or god-players, or dead people of the past, or money, or philosophy, or illusionary religion and tradition in blind-faith to create the Goldilocks' conditions with precise, scientific perfection. This is a scientific reality, that there is something that makes it possible for life to flourish! This chapter establishes the understanding that the Energy of the Universe, the energy of the cosmos or God, is in you, around you, and is always there for you, to be a free self-thinker, to create a more accurate understanding of life's reality of science, and to become a humane-being.

All religions, philosophies, and Atheist civilizations are created by leaders. They only lasted an abbreviated time; because they didn't create a better world, however, long ago free self-thinking Hebrew people in search of universal truth who understood democracy in the reality of science are still here today, even though they are a minority, because they are brilliantly scientifically creative humane-beings, and they create a better world, because people in search of new universal-truth, will only tell you the truth!

How have things changed by free self-thinking in scientific knowledge of humane-beings? Not by memorizing knowledge. For example, we came down from the trees and began to walk upright to collect food and hunt. When we needed to go faster, we domesticated horses. When we needed to transport heavy things, we built trains. When we needed to go even faster, we made cars. When we needed to go much faster and farther, we created airplanes. When we wanted to learn more about the world quickly, we invented the internet, computers, and the cellphones . All this was created by free self-thinking scientific minds and debating in a democratic country, by more accurate understanding, by not competing, and learning from others to create a creative life. It was not created by a belief or opinion or philosophies.

Albert Einstein left his family and moved in with his female cousin. It was more important to prove his theory of relativity, which proved that all energy is related to unite all the people in God's reality, for the love of life, because he was a free self-thinker, humane-being, not a **sheep in the herd of grazers.** Einstein showed by his equation of $E=MC^{2,}$ that matter can become energy and energy can become matter. Scientists that wondered about the Creator's magic of life did not think they could have done a better job. Da Vinci thought the man-body was so perfect he used it as a measurement device.

Hierarchy memorizes the illusion of knowledge, thinking they know everything. However, they don't understand the magic universe reality of what they know or will ever know because there is no understanding of the spectrum of magic of God's reality in an illusion of knowledge or in what they know in their computer materialistic brain. only by accurate understanding in the infinite mind, people and hierarchy will become a humane representative of life, because as it is, in the reality of science, today, creative people are creating a supercomputer that will have all the answers to all questions in scientific reality, to replace all the scholar leaders, and all knowledgeable books for mor accurate understanding in a debate, it will use mor electricity then the country of Israel.

Albert Einstein said: "God does not play dice." Einstein understood what he called "cosmic understanding" where God's presence was evident in the order and rationality of nature and the universe in all expressions. However, religious tradition in blind-faith is gambling addiction, and is playing 'dice; it is accepted by the medical field, as a sickness that is hard to cure, and everything that is subjective to each group. It is not possible to be the truth for that time, and everything that is not truth, is a lie because it is created by a magician's show from leaders manipulating words (god-players) for more money; and if you believe that a magic propaganda show is real, then you are ignorant. All religions and philosophical movements are created by cruel animalistic leaders who memorize knowledge, **no exceptions.**

Albert Einstein is not a philosopher or religious! Albert Einstein is a scientist, a free self-thinking Hebrew, in the search for universal truth's accurate understanding to create new knowledge.

Leon Halfon is not a philosopher or religious! Leon Halfon is a free self-thinking Hebrew in search of universal-truth's accurate understanding to unite all the people and create a utopian world.

All believers in religion create all wars that kill themselves, to be able to kill other people from other religions and kill free self-thinkers, humane-beings that understand the Creator's life Energy in reality of science, loves everything, and everybody. They are killed for not believing in stupid religions, and for not participating in competing egos that destroy life!

All religions, traditions, illusions of knowledge, and philosophies that build structure to families and the world, created by hierarchy leaders. However, all structures that are built from the illusion of knowledge are dangerous for all lives that are in the structure because the structural have to be scientifically safe. That is why intelligent free self-thinking countries have scientific building codes, and inspectors, to enforce the codes. Therefore, you have to be out of your mind to structure for yourself, your family, your friends, your country, and your world, by religion, traditions, Atheist, philosophies, beliefs, and opinions, created to protect cruel selfish hierarchy privileges! You find the correct answer within yourself.

PART ONE

What This Book is About

CHAPTER 1

FREE SELF-THINKING

My life journey proves that debating in a scientific method and searching for a universal truth for the time unites different people. It worked. It will unite all people of this planet because Universal Truth combines everything in the universe to create a more accurate understanding of Life's reality necessary for the wisdom to become a humane-being. After all, life is a living reality. **This book is about individual free self-thinkers. You are what you do, it is not about competing for ego. However, most people are part of a religion, or tradition (that is a gang) and are brainwashed by leaders of the world. I is ego, because, as a singular, you are part of everything, and that will open your horizon to love all. However, a leader's ego derives only from winning the competition.**

If people stop believing that they are important, because they belong to religions, traditions, Atheist philosophies, opinions, and hierarchies, which create blockage in your brain, called cognitive dissonance, to understanding reality, there will be no wars!

Ever since the beginning of free self-thinking, humane-beings have been and are still, against hierarchy because hierarchies created all religions that are (illusions of knowledge) to control the people that make a united "gang" willing to defend, attack, or pillage, or rape other gangs created by competition. These hierarchies created a world full of (united groups) illusionary thinking as a "norm." Therefore, the people do not make all religions. Leaders of the people create religion and tradition—no exception. Leaders must force violently, using torture chambers, to scare the people to promote illusionary thinking about themselves, by religious law, which lets them stay in charge.

All illusions of knowledge are fairytales, they are not the truth, they are lies. We currently have a world epidemic that has killed many people and caused much suffering. It will get more dangerous in the future. This shows that the hierarchy gangs, the regimes, are unable to find a solution because all solutions are solved by free self-thinking humane-beings that understand the energy in God in reality of Life, in the infinite mind. Not by competing for more knowledge in their materialistic brain! If you want to believe, you can believe anything.

That is why it is subjective to each person or group. It is time to encourage people out of believing illusions of knowledge into a scientific method of free self-thinking; by debating to learn from each other to arrive at a Universal Truth for that time, because Universal Truth is not subjective to each person or each group. All people will unite to create a love of Life, to do what is right, for a safer future. And leaders are god-players, parasitic scholars, memorizing knowledge to compete by preaching propaganda for a ranking scale. They are not humane-beings searching for the scientific debate for Universal Truth. What I mean by hierarchy is... people in government and in a position that serves the people raise their salaries or services without the permission of the public **because they can.**

This book is about respecting all the people of this world that created and are creating a better world for us! By the only way! And that is an accurate scientific understanding of the universal truth reality of that time, as we know from archeologists of enchanted stonework.

The Hebrews are not religious Jews, they are free self-thinkers who scientifically debated for accurate understanding of the spectrum of magic life reality, to solve all problems for over 4,000 years, to create a better life and a safer world for all lives. Because there is no going back to save what is extinct,

Through debate, humane-beings created the knowledge to convert all senses to sight, by developing gauges to see all senses accurately, giving people the ability to understand accurately all senses in reality. To give people the ability to build a car, train or even a rocket ship to the moon, because relying on opinions, religious beliefs or Atheism philosophy doesn't provide accurate reality to create new knowledge and that is the reason for a scientific method of debating to arrive at an accurate understanding that challenges the reader to look beyond the norm and let go of biases, beliefs, traditions, prejudices, hate, excuses (justification) that are self-destructively from being the most dangerous stupid people, by memorizing knowledge and competing. That becomes the common goal of living the norm and will become more dangerous because you are looking one way at the magic of Life, as looking at any magic show, that creates an illusion of knowledge and makes it very hard to be a free self-thinker.

If you look at the magic of Life from different directions. for a different outlook, by the scientific method of debating, by humbling yourself to learn from each other to have a spectrum of the magic of Life, it will create a more accurate understanding of the universal truth for that time, of how God's magic Life reality works.

For example, Albert Einstein's theory of relativity could not be introduced for 10 years of debating in reality, to arrive at a universal truth, because the theory was not accurate to become scientifically a universal-truth. In the future, a computer and your cell phone will tell you if the conversation is a likely scam or a belief or propaganda or an accurate scientific conversation worthy of debate. For an accurate understanding of life reality, because of blockages in the brain from illusion, it makes it very hard to know if a conversation is scientific.

My inspiring life experiences exemplify how one person can go beyond self-imposed limitations, which are just illusions that keep you from living fully because you are trapped in an illusory jail that enslaves you to your delusions. If all the people don't free self-think and debate in a scientific method to have an accurate understanding of the spectrum of magic life reality, people will end up as destructive parasite hierarchy and homeless people because they're just very intelligent animals that compete by memorizing knowledge, with no accurate understanding of reality to become humane-being.

My goal is to help you identify those illusions and then encourage you to put them aside and leave through a door that is not even locked, for your own good and all life! You are free to leave your illusions behind at any time with an accurate understanding of reality. However, delusions are always dangerous and will likely become more dangerous in the future as the world faces increasingly complex problems. Ultimately, it could threaten Life on the most beautiful planet in the known Universe. So, get back to reality. We can discover for ourselves the causes of all problems and become empowered by wisdom to solve them.

Today's political hierarchy is an obvious example of how we cannot depend on regimes to do it for us. It is essential to create fairytales to help children follow the right path towards becoming humane grownups. However, it is counterproductive to develop illusory religious and political propaganda fairytales for adults.

Wisdom is the universal truth of the time and Universal Truth is the key to accurate scientific understanding that will create the solution for all problems. **For example, the leader, King Henry VIII created a new religion for more privileges, that was to be able to kill his wife and that created fewer religious laws for his people. This is the reason English people are very creative!**

To create solutions, you need an accurate understanding of scientific reality for the changing times, by free self-thinking and debating. However, any fool can acquire a more profound knowledge of philosophy, and propaganda, by memorizing preachers' lectures in computers, or books, or by studying a religion without debating in a scientific way, and that provide a large memory of knowledge, without understanding.

THE ENEMY OF UNDERSTANDING REALITY IS CREATED BY MEMORIZING KNOWLEDGE ONLY.

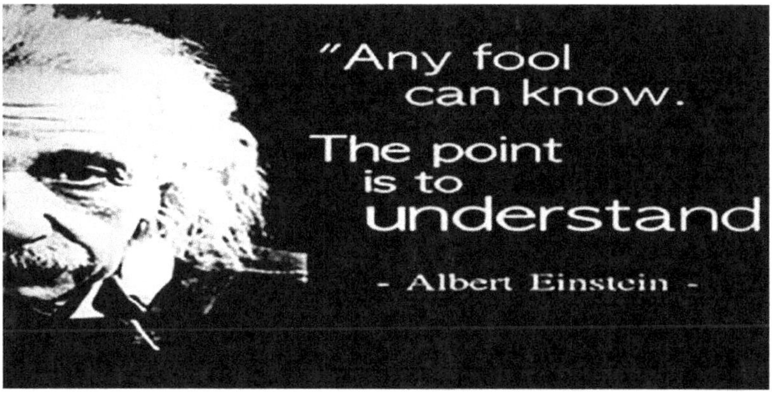

All beings are capable of memorizing knowledge. However, only people can create a more accurate understanding of reality by scientifically debating what is necessary to generate solutions by caring for humane-beings, to solve all problems. Otherwise, without a more precise understanding of reality, all animals and people are just beings because they have no accurate understanding of reality.

TWO SCIENTIFIC EXPERIMENTS

Here are two simple scientific experiments to clarify the actual reality for organized "gangs," i.e., religious followers, Atheists, or believers.

Experiment # 1. Ask a young child who hasn't been told to believe any beliefs yet, "Is there God?" The child's answer will be, "I don't know." A child is born a free self-thinker with common sense in the search of universal-truth wisdom. Therefore, it would be impossible for the child to say there is no God and I am an atheist, or this is my religion and tradition because this illusion is created by preaching propaganda, not reality!

Experiment # 2. Teach a child this fundamental law of scientific physics, "For every action, there is an equal and accurate opposite reaction." Then ask the question, "Is there God?" The child will answer, "If there is life, then there must be something that started the action to create life. Therefore, life is a reaction. The "action" that made life possible is what people call the creator, God." This is a scientific free self-thinker's answer. We still don't know how and why anything in the Universe got here. All we know is that energy is in everything, and everything is created by energy, all is in one, and one is in all.

CHAPTER 2

TRADITIONS

All Tradition was not created by the energy God, or the 10 commandments, or the American Constitution! Traditions are long-established customs and beliefs from dead people of the past. People who believe in traditions and other illusion fairytales become slaves to their traditions because they like to feel a part of something big, and **that makes you believe you are important**. But celebrating them is not essential because they repeat the same celebration from generation to generation that keeps your idol, and it separates people from different traditions.

However, it is more challenging for free self-thinkers, more fun, more inspiring, and more important to be part of everything, which is more significant than just something big, and that is democracy for all people to free self-think to be creative for a better life. What are traditions? Slavery is a tradition, religion, atheism, royalty, hierarchy, preaching, bribery, dictatorships, gang, believing, competition, superstitions, bullies, caste systems, war, and peasants to be used in blind-faith by leaders' hierarchy.

Wouldn't it be wonderful if there was a theme park where people could visit the world's traditions and religions in one place to discover the foolishness of all beliefs and practices? The people will be amused, compare, and understand how ridiculous beliefs and traditions are without traveling. As the government, a representative of the people can create this amusement park to help people become free self-thinkers with a more accurate understanding of life's magic reality. So, people can be free self-thinkers and creative humane-beings to save this world from extinction! It is time to create a brand new one-holiday celebration worldwide called the Festival of Life.

Every January eleventh: One-one-one (1/11). All life as one, to rejoice, which unites people and animals to remind us that we need each other because we are all in this world together and in life, only universal truth for the time exists. Religion and traditions are **nothing or 0** zero. Free self-thinking in search of universal truth to create new wisdom from reality is a scientific method that always works; and holidays should not be religious, because it separates people. It is time to change for a democratic world, and have time off from work as a holiday, just as we have time off from work every week and time off work as a sick leave.

CHAPTER 3

WISDOM

Wisdom does not come from belief. Wisdom comes from an accurate self-understanding of reality. The lyrics of the Beatles song "Let It Be" derive from the religious concept of accepting one's blind-faith and *in times of trouble faith comes to me, speaking words of wisdom from belief, let it Be.* But the Beatles and most people look for wisdom in religious beliefs; that is the wrong place because that is what they have been led to believe by preachers. Let it becomes from god players. However, God doesn't play dice. His presence was evident in the order and rational understanding reality of nature and the universe in all lives.

Wisdom comes from more accurate self-understanding by extensively debating subjects that have occupied the greatest minds in history. Their wisdom did not come from doing nothing, just "letting it be" or blindly in blind-faith addiction, following leaders in charge. However, many free self-thinkers were punished for making a better world throughout history because they were minorities, and cruel hierarchy, and lazy followers who were ignorant of reality who were the majority. Perhaps they should have "let it be," but they didn't, because the energy of the universe is in you to think freely, which is the right thing. Instead, they sacrificed their lives because they did not have tradition from beliefs and understood right from wrong by the creator God's reality of science to create new universal-truth of that time for a better life for all lives. However, today, most people are free self-thinkers, not stinkers, from understanding Moses, understanding the American Constitution, and understanding democracy.

The Beatle, Sir Paul McCartney, accepted knighthood as a tradition and became hierarchy, that is not wisdom. Steven Hawking, who understood scientific reality, did not accept knighthood, or become hierarchy, that is wisdom in a democratic reality of science for all lives.

RELIGIOUS JOKES
LOOKING IN THE WRONG PLACES

One night, a man was searching underneath a lamp post for something. People passing by asked, "What are you looking for? The man replied, "I lost $100." They asked, "Where did you lose it?" By pointing, the man responded, "Over there in the dark." The people asked him, "Why don't you look where you lost it? The man said, "You think I am a fool? I cannot see in the dark, but I **believe** I can find it under the light of this lamppost." **"Because if you want to believe, you can believe just about anything."**

Have you heard the story about a religious woman who fell in love with a scientist? After their marriage, they had a baby boy. When the baby was older, he asked his mother, "Where do we come from?" The mother answered, "We come from Adam and Eve, sweetheart." The boy responded, "But mom, dad says we come from apes!" The mother said, "Your father is right. Monkeys come from his side of the family."

A man died and met Saint Peter in Heaven; Saint Peter said, "No, no, no, what you did, is not a sin, hate, is an ignorance of God in reality of science. You just worried yourself to death over something that was not sacrilegious." The man said, "How should I know? I am just a believer of leaders, in blind-faith, not a free self-thinker in search of universal truth!"

A religious leader places his hand on a young man's head and gives him a blessing. Then the young man puts his hand on the religious leader and gives him a blessing. Next, the religious leader extends his hand and asks the young man for $5. The young man extends his hand and asks for $10, saying, "I usually ask for $20, but I am giving you a special religious leader discount." Finally, the religious leader says, "I don't believe you." The young man responds, "I don't believe you either because I am not a believer; however, I understand realty by free self-thinking and debating.

A devoutly religious man enters a cab in New York City and asks the driver to turn off the radio because his religion bans him from listening to infidel music. The driver stops the cab, opens the passenger door, and says, "Back in the time your religion was created, there were no automobiles, only carts and wagons. Cars, trains, and planes were made by inventive free self-thinkers searching for universal truth, not delusionary beliefs and lies. However, if you still want to believe, maybe you should get out of the cab now and wait for a flying carpet to take you where you want to go."

A farmer sold his donkey for $100, to a leader. The leader requested that the farmer keep the donkey for two days. The next day the donkey died. The farmer finds out that the leader made $1000 from a dead donkey, so the leader explained that he sold 501 raffle tickets, for $2 each and returned the $2 to the winner because the donkey died!

A successful religious young man brought many young ladies home. However, the mother disapproved of all the ladies, so he asked a friend what to do. The friend said, "Why don't you find a lady that looks like your mother." **He did just that; however, his father disapproved of her!**

A religious man and his partner had a religious dispute, so, they went to see a Babylonian Rabbi to solve the dispute. The Rabbi, after listening to each man's dispute, said to one man, "You are right," and he said to the second man, "You are also right." The two men responded by saying, "How is it possible for the two of us to be right?" The Rabbi said, Religion is an illusionary belief, and if you want to believe, you can believe just about anything, and that is the reason there are many beliefs, because religious belief offers no confidence, no certainty, and no accurate understanding.

CHAPTER 4

THE SOCRATIC METHOD

ALL ANSWERS Are in THE QUESTIONS for the search of universal truth. FOR EVERY QUESTION THERE IS AN OPPOSITE ANSWER **for a more accurate understanding.**

- **What is darkness? = Darkness is the absence of light!**
- **What is a belief? = accepting what someone says!**
- **What is religion? = the absence of the creator God.**
- **What is propaganda?= the absence of truth!**
- **What is war? = War is the absence of humane-beings!**
- **What is in blind-faith? = the absence to see the light.**
- **What is hierarchy? = the absence of representatives.**
- **What is an obedient follower? = the absence of democratic creative free self-thinkers!**
- **What is ignorance? = ignoring reality!**
- **What is Atheist philosophy?= the absence of understanding Scientific reality from debating.**
- **What is tradition? = the absence of continuous change.**
- **What are illusions of knowledge? = the absence of understanding reality to solve all problems in life.**
- **What makes you immoral? = the absence of understanding in science that there is a creator!**

No one can change anyone in life, no matter how good the intention or wise approach is. However, in a debating conversation, you can let others say what they believe. Then ask questions to reveal the contradictions in their thinking, to ask themselves important questions and eventually arrive at an accurate understanding. Encourage them to do the reasoning to understand for themselves, to change their life and the world for the better from having confidence in God in reality of science! A believer is someone who believes they know everything; however, they don't understand the magic universe truth reality of what they know or will ever know because believing is an illusionary knowledge. There is no understanding of the reality in what they know because religion is a gang common-belief, not common sense!

The Socratic Method examines an assumption by asking questions about why it is believed to be the truth. Ask where the evidence is for it and what are the alternatives. You will arrive at a more transparent, more accurate understanding and be less ignorant by doing just that. That is what free self-thinking in the search for universal truth is all about. **However, competing questions, it is not a Socratic Method, it is an animalistic competition that pokes holes in someone's life picture and only sees the holes, not the picture!**

In ancient times, the philosopher, Socrates, was taken to a Greek court and prosecuted by Meletus and two others who were members of the democratic elite (hierarchy). They found him guilty on trumped-up charges of impiety and corrupting the youths of the country. Socrates used his scientific method of thinking to show he would have no reason to do what he was charged with. Nonetheless, his jury of 501 citizens found him guilty, **even though they loved Socrates, for his devotion to democracy**.

Because of their greed for comfort, of money, they must have their children believe in the illusion of knowledge and tradition; to believe they are important Greeks, because of the need to be important and that is a universal-truth. Socrates was ordered to suggest his suitable punishment. He replied that he should be sentenced to dine in the Prytaneum, a state honor generally offered only to victorious famous generals, because he understood that he created a better world and a safer Greece. When his "punishment" was denied, Socrates suggested death as a penalty, to create a more accurate understanding of hierarchy **that promotes freedom of belief and tradition that is the illusion of knowledge and to believe in blind-faith that they are** important. **However, in democracy, nobody is** important **from belief, and that is the reason in history all democracy was and will eventually be destroyed in the future by freedom of religion and tradition because it does not work together.**

He was offered a chance to save himself if he was exiled to a desert island (Djerba) for the rest of his life, but this was unacceptable to Socrates. He wanted to make a point about the wisdom of his life and prove that if this travesty of justice transpired in democratic Athens, similar injustices could happen in any democracy. In the end, Socrates became a martyr for his cause and was universally renowned. Socrates was a great free self-thinker searching for universal truth to find wisdom, a Lucifer "Satan," who enlightened and changed the world for the better. He should have suggested being exiled to the island of Djerba, where he would have been appreciated.

Do you think Socrates deserved death for teaching his students to think for themselves by asking themselves questions? The Socratic Method revealed that an Emperor is not a God, that religious leaders manipulate words, and that believing is neither knowledge, nor wisdom, nor reality, nor a universal-truth. The Athenian nobles claimed that the Socratic Method was corrupting the youth because it revealed the truth. This is a lesson about religion and hierarchy. Those in power must change to become representative champions of the people or become extinct, which is just what happens. Eventually, greed from hierarchy became extinct, and destroyed the great democratic civilization of Greece, and it will destroy the great democratic civilization of America, if we don't, stop freedom of the illusion of knowledge!

If you get attacked by a harmful germ or virus, you will survive, because you love life, and that makes your energy stronger. If you get attacked by bullies and survive, you may decide it is necessary to become stronger, and smarter and learn how to defend yourself. You do it because you don't want this to happen again to you or other humane-beings.

People should stay healthy by exercising their minds in the reality of science to understand how gravity affects their bodies. You are constructed on the frame of a skeleton. Muscles and tendons keep your bones and organs in the right place. If you don't exercise and stay comfortable, gravity will pull the bones against each other, creating pain. To remain healthy, you must understand the importance of challenging yourself by exercising your body and mind, as you must know that life is not about comfort. Throughout history, religious traditional countries that are idle, by following and obeying an aggressive leader, lost the war to countries that exercised their minds to create powerful weapons!

CHAPTER 5

MASONS FROM THE CRUSADERS

The Pope recorded the Knights to protect Jerusalem, the home of Jesus. The Knights were renamed Crusaders by the Pope, and they preached to hate the free self-thinkers, the Hebrews. However, when occupying Jerusalem, the Crusaders searched for the object from history and found the King Solomon Book of Masons, which is Hebrews thinking in God's reality, **"You are what you do, not what you believe you are."** So, the Crusaders became free self-thinkers, creative Hebrew Masons, who created over 8,000 of the most beautiful buildings in Europe, and around the world, and started banking, to help free self-thinking creative people. However, the **Babylonian religious** Hierarchy, and King of France didn't want free self-thinker creative people. They wanted ignorant obedient religious followers of tradition in blind-faith. So, the Hebrew free self-thinker, Mason Knight Templar, was burned alive at the stake; and on Friday the 13th, the religious army massacred many Masons **because they were a minority**, to keep people as ignorant followers in blind-faith. So, ignorant leaders can stay in charge.

So please don't tell me that religion makes people moral.

This ignorance movement repeats itself today, and it will stop only when the majority of people free self-think and debate. The Masons escaped to America and declared war on the parasite Hierarchy. After winning the war, the Masons wrote the Declaration of Independence and the American Constitution by the people, for the people, from the people. No religion in was government allowed, creating a better understanding of life for the people in the world, because life needed a humane-being who was a representative for all the people. To have the freedom to think in the mind and become a creative humane-being, only then will there not be a problem that cannot be solved.

If the founding fathers, were opinionated or believers, they could not have written such a humane American Constitution. The saying, "Sticks and stones will break my bones, but words will never hurt me." This saying is to protect the right of free debate and conversation in scientific reality to create a more accurate understanding. The Masons became a secret society for debating in the search of God in reality of science, because religious leaders persecuted them, and they were a minority of free self-thinker humane-beings; cruel leaders did it and are still doing it because they are not humble enough to learn from others. The Masons understood that there was a creator; the energy of the universe and the 10 commandments are in the American Supreme Court.

The reason the Masons are not elected as president of this country is because the Masons hired assassins to kill religious leaders to create a free self-thinking country in God's scientific reality, because democracy unites all the people, and a parasite Hierarchy creates a gang from religion that separates people, and that is against democracy.

The Masons did not allow women of that time, in the men's lodge for debating in scientific reality, and they were not allowed to vote, because women of that time were not thinkers. It was a problem of that time, today, all men must free self-think and debate to protect democracy, and women who want to, are capable of being brilliant to protect democracy, by changing with the time. **However, women have a biologic time to be a mother. So, free self-think, because it is your life!**

CHAPTER 6

UNITED STATES OF AMERICA: THE GREATEST COUNTRY IN THE WORLD

The American Revolution rid the colonies of gangs and the British Empire's hierarchy, and that energized the individual American free-self-thinker's search in reality for universal-truth wisdom. America's founding fathers debated in scientific reality for four months and by free self-thinking, they arrived at the famous compromises leading to the Declaration of Independence Constitution. After lengthy deliberations, wise old Benjamin Franklin said, "I am surprised that all this disagreement can arrive at one answer." But somehow, these free self-thinkers pulled it off by finding God's universal truths of that time. Had the delegates' debates at the Constitutional Convention failed to reach the "one answer," that would mean they had been unable to achieve universal-truth wisdom for that time.

> *"Christianity is the most perverted system that ever shone on man."-Thomas Jefferson*
>
> *"I have found Christian dogma unintelligible." -Benjamin Franklin*
>
> *"This would be the best of all possible worlds if there were no religion in it!"-John Adams*
>
> *"The Constitution is the guide which I never will abandon."-George Washington.*
>
> *"All religions were inventions set up to terrify and enslave humankind." -Thomas Paine*

The American humane Constitution separated religion and hierarchical gangs from the government. It was the second constitution of humane law founded without establishing a national religion or belief for parasite hierarchy. But, on the other hand, the founding fathers in Philadelphia designed the United States to be self-governed by free self-thinkers' representatives by the people, for the people, from the people.

Although little acknowledgment in modern times is that the first democratic humane constitution was Moses' law between God and people, to create free self-thinkers. It was founded to understand God as life energy. Moses wanted to unite all the people of the world in God in reality of science. without religion for god-players. Moses' objective was to create a community based on debating by free self-thinking to search for universal truth to create continuously new wisdom. Thus, the Ten Commandments were not intended to become part of religious dogma but rather a sacred constitution between God and all people, to create free self-thinkers, humane-beings, out of people in a democracy. Not for religion, and tradition for parasite leaders (god-players)!

America's founding fathers were all Masons. The ancients lived by the Book of the Masons, written by King Solomon in Hebrew, to unite people about understanding the God of the Cosmos. He did not create a religion, but he did send missionaries to enlighten others in wisdom. As a result, America became an example of morality, creativity, and fairness for the rest of the world. It opened the door for the opportunity to unite free self - thinkers, across the globe. As a result, many people from all over the world want to move to America. They want to be allowed to free self-think. They come here to unite free self-thinkers through wisdom and scientific knowledge for a more accurate understanding of reality. However, the Masons did not find the book of **Adam, the snake, and Eve**, to debate for a more accurate understanding of the meaning, to have the wisdom to not underestimate the slippery snake hierarchy leaders; and that created an opportunity for the sneaky snake parasite hierarchy to get back. So, the parasite hierarchy created a place to listen to ignorant conversations and **preach the illusion of knowledge and religious forgiveness. It was called** a **confession room** and that created a large movement of religious followers of women, who gave the hierarchy the power and privileges from illusionary knowledge, religion, and tradition. **Today, Atheists are changing history books by saying that the founding fathers were Atheists. That is how hierarchy creates propaganda and lies, because most people cannot understand how you can exist without believing in something. However, when you are searching for universal-truth in a debate, if someone says what they believe then the search for universal-truth cannot continue!**

CHAPTER 7

THE MOST CREATIVE COUNTRY IN THE WORLD USED TO BE THE UNITED STATES OF AMERICA

America had bars everywhere and everybody knew your name from debating conversations and discussions that created the wisdom of the American Revolution and evolution that led to the wisdom of the American Constitution: By the people, for the people, from the people, not for hierarchy. Two hundred years later came prohibition, which outlawed and banned all alcoholic beverages from the country**. This unenforceable law was created by a large movement of obedient religious women believers in blind-faith, and that destroyed those debate conversations**. Banning alcohol ultimately led to organized crime, such as bootleggers, and rum runners by the Mafia, which created powerful police, and powerful judicial hierarchies that arrested innocent people for drinking and started a decades-long war between criminal gangs and police gangs, with innocent creative civilians caught in the crossfire. It was created by ignorant obedient followers of religion's illusion of hierarchy greed. That created religious law that created power hierarchy laws, **The wave of that religious movement still exists today, and bars are mainly meant for getting intoxicated, not necessarily used for debate conversation. As a result, today** America is a Christian country and Christmas is a national holiday, **to protect hierarchy** business for more privileges. It is no longer a free self-thinking democracy; it is a republic.

Gangs will always win if people have no power for free self-thinking because free self-thinking creative people fear too many laws that they don't understand. They need a hierarchical lawyer and a hierarchical judge to keep them out of jail, and that is the reason today, there is a lawyer for every 100 Americans.

The representative of the people in the government's job represents all the people to level the playing field. So, the government created by the people creates wisdom by free self-thinking, which is necessary to develop confidence and to welcome different views outside the norm!

Some people are raised to be believers that they think are important by following tradition and become an aggressive inferior gang, that have to be aggressive to fill important, while others are raised to think for themselves and that is why all people cannot be friends. But, on the other hand, all people need scientific thinking to be friends. It is wisdom for this time, to include all people in a more accurate understanding to arrive at God's universal truth reality of science because all people are capable of being brilliant free self-thinkers for a utopian world!

What matters to all great free self-thinkers is what they think of themselves, not what other people think of them from illusion. Then, by judging themselves after debate conversations, they will understand what is right and realize that humane-beings don't have to compete to impress anybody. The Socratic Method was intended to improve the understanding of reality, not to compete. Plato's "Republic" book is about the most prolonged debate between Socrates and his friends. Plato speaks in the voice of Socrates, who is the central character in the book. However, most of Plato's books were destroyed by religious leaders to enforce religious laws to create ignorant illusionary believers in blind-faith addiction,

Orville and Wilbur Wright often debated with each other in the search for how to design the first motorized flying machine. After two years of scientific debating, frustration, and failure, the brothers finally figured out how to create a propeller that would work. Each blade at both ends was in the shape of a boat sail, so the propeller could catch the wind to pull the plane forward. Without an accurate understanding of God in reality of science, they could not have designed a powered aircraft that would fly. They created the wisdom of a flying plane's propeller because of their scientific debating, and it is still used today because they reached universal truth reality of the time, which creates new knowledge.

Today, many corporations, interest groups, lobbyists, and politicians are gangs in America. They are chipping away at the people's rights by weakening the Constitution to inhibit freedom of speech from greed because freedom of speech can create disagreement. They want to control the country via biased media, digital disinformation, and so-called "opinionated news" shows on radio and TV that create and promote half-truths and freedom of religion in the country, to create ignorant obedient followers for more votes to control the people, because they have your money to pay for it. However, the news must debate with people from the public to arrive at a more accurate understanding of the information. Many people think it is time for a new evolution, not a revolution, because it is no longer a free self-thinking democracy; it is a republic.

They are looking for brilliant wisdom that will unite humane-beings, so they can separate illusions from scientific wisdom, create a government by the people, for the people, and encourage free self-thinking in search of new wisdom that brings out the best of humane-beings.

We can secure a better future not just by reading, but by questioning what we read; not only listening, but questioning what we hear; not just watching, but questioning what we see and not just believing but questioning those beliefs; and in a debate ask, "is this a belief, an opinion, a theory, or a scientific conclusion?" In a debate, challenge yourself and ask how anything was discovered or created for a more accurate understanding, which will inspire the people of the world, and that will create a majority of free self-thinkers, to create a democratic world!

CHAPTER 8

DISCOVER FREE SELF-THINKING IN SEARCH OF UNIVERSAL-TRUTH

Over the years, it has been determined that many health problems come from the medical model of relieving symptoms rather than curing diseases. This has led to a culture of patching everything in life—as leaders would have you do to maintain power and control. Chronically ill patients who return for more treatment are far more profitable than healthy patients. It is suitable for business because of greed from competing scholars who think they are entitled to make more money, by using and abusing people.

WISDOM creates humane-beings and wisdom will give medical science the power to be creative with natural cures. However, leaders want to use whatever they can, including robots, to control the people. But the future is not robots or knowledge. It is **WISDOM from humane-beings.**

A believer, a gambler, or a follower, all rely on chance or superstition, for God to be responsible for the outcome. The question is, what are illusory people doing on this planet? Is it to overpopulate, destroy the environment, let everybody believe in whatever they want to believe, for hierarchy control by keeping people from free self-thinking and twiddle their thumbs? However, a free self-thinker searching for universal-truth clearly understands that people must become humane-beings to care. They do not expect chance, superstition, or god-players to do it. They rely on wisdom from God in reality of science in search of universal truth because life is an inspiring journey of challenges, not a monopoly, or a monotonous comfort journey.

Wisdom from Albert Einstein: "Imagination is more important than knowledge." This imagination must come from free self-thinking. All other sources have no universal truth. When Einstein wrote his theses, people told him, "What you are saying is that Newton's wisdom was wrong." Einstein replied, "His wisdom was right for his time and my wisdom is right for my time." Because life is not a competition for humane-beings, wisdom changes with the changing times, and universal truth changes with the time. Our planet belongs to all living things. Nobody has the right to use the land, seas, resources, and overpopulation like money in the bank by the illusion of hierarchy greed. They must be shared for all life to flourish. That's the only road for people to evolve into humane-beings because nobody purchased the planet Earth from God, the Creator of life Energy of the Universe! This is one reason we pay taxes. Hopefully, some of that tax money is used to preserve, protect, and maintain our Goldilocks' atmosphere for all life to flourish from the wisdom, by debating to create an accurate understanding.

Albert Einstein thought that mathematical equations could explain the whole universe. But he was wrong because the universe continuously changes. Mathematics is Knowledge and knowledge does not change, only wisdom changes Mathematics and Knowledge. Therefore, today's wisdom will not be the wisdom of tomorrow!

Science searches for one trustful answer at a time. Every brain lives only once in one life. Every man, woman, or child is entitled to enjoy the individual's right to the most important right. And any individual free self-thinker in the search of universal-truth is capable of being brilliant.

As humane-beings create wisdom for a more accurate understanding of the world, the more the world becomes part of humane-beings and when the humane-beings become part of the world as a whole, it creates symmetry. That is why humane-beings don't want power over others by competition, violence, wars, greed, or propaganda. In the end, there are universal truths for the time. However, we will always be things, and we will never know everything, and are not supposed to know; only having a real understanding in your wisdom will be creative.

When you become a humane-being, any effort to maintain an unsustainable status quo or use power over other people disappears. Leave behind the alienating world of hierarchy and the idolization of leaders and tradition.

Since the beginning of history, there has been vast knowledge for advancing a better life for all lives by free self-thinkers. However, most of the knowledge and free self-thinkers were destroyed by hierarchies that are lying **to stay in charge**. Free self-thinkers have been tortured or killed by leaders and still do today by gangs of ignorant followers who believe in illusory knowledge.

The greatest enemy of knowledge is not ignorance, it is the illusion of knowledge, and that is a magic show created by religion, tradition that repeats, from generation to generation, and does not facilitate change, and that will keep you idle. Only by free self-thinking in a debate, you will create new knowledge!

For example, after observing a herd of grazers, the lion goes after the logical choice, and that is common sense; the weakest animal he can catch and eat is logic. Most animals and homo sapiens belong to a group, to be safe. That is logical. Thinking that animals have no logic, feelings, or intelligence comes from religion, beliefs, and illusions of knowledge rather than scientific truth. Leaders create a hierarchy by telling you that you become exceptionally important when following them in their illusion of knowledge. But the truth is, you are what you do and if you are not humane, you are not a human or a humane-being you are just a being! In addition to logic and feelings, people have higher intelligence from memorizing knowledge, thus enabling us to become the most powerfully destructive animals.

That is why leaders of people and animals were and will not be able to achieve the universal truth of that time. Because they are not humane-being. All People are equally capable of maturing into a humane-beings. By free self-thinking and in a scientific debate conversation, they can overcome obstacles of the brain from believing in the illusion of knowledge. Science deals with the provable reality; this is why humane-beings can achieve wisdom by correlating with all living things to achieve universal truth for that time. The universal truth is the language of the universe. If it is not simple, does not correlate to a universal truth for that time, and does not unite all the people, it is not God's reality of science, and it is a lie. In this stage of evolution, nature does not need people. However, nature has a desperate need for humane-beings! Humane-beings have learned to defend themselves powerfully because they are creative. Civilization does not need leaders, but civilization has a desperate need for humane-beings to act on behalf of all life for a safer life. Then people can indeed be independent, free self-thinkers to discover more new knowledge.

Why do you believe that all homo sapiens are humans? Why do some civilizations believe that animals are gods or that wolves are people who came back to earth to teach us how to hunt and survive, or that rats are reincarnated ancestors? Many civilizations falsely believed that nature exists only to use and abuse. Where is the evidence? Look at the mighty peoples of the past: The Egyptians, the ancient Greeks, and the Romans. Their populace was taught to feel important by becoming followers of the regime. However, if people are not humane-beings, they are merely animals; and all these great civilizations and traditions disappear.

How do you know that some people are humane-beings and are not just animals? All animals need a "carrot at the end of a stick" to propel them to act, such as hunger drives hunting for prey, grazing for food, or protecting their territory. They may also work because someone is whipping them with a stick. Carrots also drive people on sticks, such as fame, fortune, or affiliation with a gang by competing and obeying religious laws. Many are motivated by the promise of an illusory afterlife, whereby followers who behave are rewarded in heaven. In contrast, bad action and non-believers are condemned to a fiery pit, eternal torment, or enslaving people who will act because someone is whipping them, by creating a scary god player!

Humane-beings don't need incentives like a carrot on a stick to accomplish worthwhile humane activities. They act to do the humane thing. They would risk their lives to save a total stranger. That is because a more accurate understanding makes them love all lives and live humanely**, to be able to understand the Energy of the Universe, the cosmos, and God because the creator is to be understood, not to be memorized!**

Humane-beings get gratification from making a better world. And doing what you know is right is gratifying because it results in the joy and love of life. Here is one example of a humane-being: After the Armenian genocide in 1915, which killed or expelled one and a half million people from the Turkish Ottoman Empire, a Danish woman named Jacobson organized and cared for 1,500 orphaned Armenian children and did not create any children of her own. Jacobson was a free self-thinking humane-being. She did it for her love of life. She understood in reality, that there was life energy, the creator of the Universe.

CHAPTER 9

EXPERIMENT TO KNOW THE PERCENTAGE OF PEOPLE WHO ARE HUMANE-BEINGS

An experiment was conducted in 1960 in which a doctor asked participants to raise the level of pain on a device that would electrically shock someone in the next room with a glass door. The participants were unaware that the person being "electrically shocked" was an actor. Every time the participants would raise the (nonexistent) voltage, they could see and hear the actor scream and exhibit more pain. The actor would even come to the point where he pretended to die if the voltage was raised to its maximum. The participants simply had to choose either to obey or disobey, increasing the purported voltage, or by free self-thinking in wisdom, not to obey by using the word "no" to a doctor's, or a professor's, or a "leader's orders!" The 1960 experiment results showed that 60% obeyed, thinking they were hurting someone and 40% disobeyed.

In 2011, the same experiment was performed and an overwhelming 75% obeyed orders. This shows that people are more likely to choose to be followers and were more obedient in 2011 than in 1960. In 1960, there was a hippie movement to free self-think. In 2011, we did not have a free self-thinking movement because of too many laws, and freedom of religions. However, we have more preaching on television propaganda, giving the hierarchy more power over people, staying in control, and going to war or using the people for more privileges.

The people on this planet were better yesterday than today. There are fewer humane-beings today and more people live by an animalistic competing law because it is the law of hierarchy. The researchers concluded that people in 2011 are busy watching professors that memorize knowledge on computers, televisions, and cell phones, including YouTube, preaching opinionated, propaganda, and inflation that create selfishness to feel important, instead of having a debate conversation like the hippie movement, to create a more accurate self-understanding right from wrong, to do what is right.

I created about 20 inventions and not one person who copied my work offered me a profit share! The inventor of the television was a farmer. He had a patent, and he showed his invention to General Electric. G.E. kept his designs until his patent time ran out, then they started making and selling televisions. The inventor farmer died poor. His life shows that hierarchy is greedy, and they are not humane-beings! Humane-beings searching for new wisdom created all great civilizations, but all great civilizations are destroyed by atheist hierarchy greed because their God is $ money!

A scientist claimed that free self-thinking for more power and conversation searching for wisdom creates people who exercise their minds and make their brains healthier, which has been shown to prevent Alzheimer's disease and help people live longer. Alzheimer's is killing more people every year and is one of the most expensive and heartbreaking sicknesses to endure. Alzheimer's disease is becoming an epidemic and there is no medical cure for it because religion, belief in blind-faith, and illusions ignoring reality, create blockage in your brain, and that creates Alzheimer's. The only remedy is free self-thinking and debate in God's reality of science to keep your mind exercised to understand the magic of life. *Imagine that?* Because the cosmos will not have it any other way! So, start debating to understand that there is a creator,

In this time of our evolution, I am right until proven wrong. Some people ask me, "How can you talk about all these subjects: religion, politics, and feelings without getting mad?" (Getting mad is getting crazy). I answer, "When you search for wisdom, you don't get mad, only people that believe in religious lies get mad, and are dangerous! First, listen to what is being said, then free self-think and ask questions to acquire a more accurate meaning of words. If I don't answer, I will excuse myself and get back to the conversation later. When I find the answer, I will respond."

CHAPTER 10

WHAT WE KNOW NOW FROM HIGHER CONSCIOUSNESS

Early homo sapiens synchronized their actions in unison gangs. This created a need for leaders and followers to work together for safety and survival. This developed hierarchies. It is as if they are done by one brain with power over all other minds in the gang; and they eventually began believing in a man as a king-god and a descendant of the God of all Gods. Belief creates duality, not symmetry. If you believe in a man in the sky who is a God, you must also believe in a man in the earth who is a Devil.

When you create a belief in religion, you also create a belief in atheism. Enchant beliefs are evil and dangerous. However, people by starting and debating in reality, create accurate meanings of words to understand life. This leads to the wisdom that transforms people into humane-beings. And as life continuously changes, humane-beings search for new universal truth and understanding, to create the wisdom for new knowledge. Wisdom must also continuously change to overcome planet Earth's existential challenge of global warming due to pollution and overpopulation. Coincidentally, the energy particle was discovered when we had the knowledge and technology to do so. Thus, everything has been changing from the beginning, continues to change, and creates exciting challenges to protect the Goldilocks' atmosphere for life.

When you look at any magic, you have to look at the magic by debating to see the spectrum for an accurate understanding. Since we can acknowledge others and ourselves around us, with a more precise understanding of creating a higher level of consciousness than another being on the planet, it is essential to free oneself from thinking in illusion to develop into higher consciousness by searching in universal truth to understand the spectrum of the Creator in scientific debate.

What is the difference between people of higher consciousness and being? We can send people to the moon. A dolphin, on the other hand, cannot send another dolphin onto land. So, what is it that makes us such brilliantly self-aware? Why are we capable of understanding things like God's universal truths of science and why are we capable of having this "higher consciousness?" Because we are here on earth for an inspiring purpose and brilliant enough to realize in the spectrum of the magic of life that creates the wisdom to know **that we are responsible for protecting the environment and all life. Not because it is easy, but because it is a hard, inspiring, and exciting challenge. That's what my wisdom has shown me during my 82 years on earth life.**

Some knowledgeable people with no wisdom theorize that consciousness is an illusion created by chemicals in the brain. They can see where thoughts occur in the brain using Magnetic Resonance Image scans (MRI) and can even estimate what you think. Still, they cannot tell where you are or where your mind is; although they have figured out where activity in the brain operates, scholars do not know who or what is in charge. Essentially, they cannot find "you or your mind" in the brain! On the other hand, as free self-thinkers in wisdom, realize that there are things that people are not able to measure or know.

However, only by an accurate understanding of the spectrum of the magic of life, you create understanding! Because if scholars are able to memorize the knowledge, they will compete to create a parasite hierarchy. Therefore, they are not humane-beings, they are just beings, and all beings don't and will not understand the magic of life, because **people who memorize knowledge and compete to become parasite hierarchy, will not comprehend infinity in their Brain**, because the brain is an organ, and all organs die. However, a free self-thinker in search of God's universal truth of science, understands that universal truth continually changes with time to create an infinitely exciting and challenging life, and that understanding in your mind creates a higher consciousness humane-being, who can understand that you are infinite. Because the energy of the universe will not have it any other way, the reality of the universe we live in has energy throughout, which provides all the necessary resources **to fix itself or destroy it!** Since we are aware of this and brilliant enough to understand much about the universe's mechanics, we can harness its energy to change our lives and all lives for the better—by wisdom from humane-beings of Higher consciousness!

CHAPTER 11

EXAMPLE OF FREE SELF-THINKING IN A DEBATE

If you lose your keys or some other essential item, you might ask a friend to help you find it. You ask them because you know two heads are better than one. That is the principle of the scientific method of debating. But, by free self-thinking, studying the situation and analyzing all known facts, by debating, science will take you in the direction towards finding yourself closer to universal truths for that time.

Throughout history, great story writers wrote books such as the Bible, the Tora, the Koran, Buddha's books, and Shakespeare books. Over time, leaders translated these books differently, changed these books and destroyed books to create illusion, for power over a believer to believe that these books were inspired by God, even written by God and their translations spurred new religions, dogmas, cognitive dissonances, prejudices, hatred, and wars. But if you have been taught to free self-think in wisdom, it will become evident that words are not perfect, have many meanings and are written by people who sometimes make mistakes and are lying from greed. God did not write these books, people did.

Religious leaders' tactics scams are copied by politicians, salespeople, and gangsters for-profit by scamming the believers, by believing greedy hierarchy. That is why we must convert all religious buildings into temples of wisdom. Instead of sending children to school to simply memorize knowledge, it is better to make them into humane-beings, to be creative by wisdom for a better world, because today all knowledge is available on the internet. It was created by a free self-thinker in search of the universal-truth of that time by debating the reality of science.

Everything in this book is open to debate. You don't have to agree with me. But, if you disagree, ask questions for a more accurate understanding of the meaning. **For example, religious believers who have read my books often say, "This your belief, and I have my belief," or "This is your**

opinion, and I have my opinion," because religious people cannot understand, how you can exist, without believing or having an opinion. However, in reality you can have understanding without believing or having an opinion.

A one hundred ten-year-old man was interviewed and asked why are you so healthy, wealthy, look so good, and live so long? The older man said, "As a young man, I understood that it is my life, so I became a free self-thinker and enjoyed life peacefully. However, after I got married, my wife started demanding and was no longer my partner, so I got rid of her. Later on, I got rid of all the parasites in my life and only kept humane-beings, because it is not wise to keep parasites in your life, or in your world.

In this book, you will not find the phrase, where I say "I believe or have an opinion" because it is not valid, helpful, or necessary and it is harmful toward creating confidence. To live your life fully, you need confidence, from understanding reality. I wrote this book partly because of my experiences, which gave me a more accurate understanding of those boxes of belief. I realized that these boxes fall short of depicting a precise perception of life. Believing that your box is better than other boxes is an illusion because every religion or belief cannot be better than others' beliefs. To gain wisdom about reality, you have to evolve with an open mind.

What if all people on this planet free self-think in wisdom, focusing on the same question? All the answers eventually will correlate to one scientific universal truth. In science, that is called finding the consensus solution that unites all responses to one universal truth of that time. All people can only focus on one thing at a time. That's why magicians can trick us with their dexterous, slight-of-hand illusions. However, people know a magic show is just entertainment. When you get out of your box and look at the magician from all directions, by debating, you will understand that God-realities cannot be seen from one direction that is known to exist as a norm, nonetheless. That's the rabbit jumping out of the hat.

Religions and gangs create illusory boxes they believe are the best. But when you are in any of these boxes, what you see is a manufactured illusion of knowledge from only one perspective or one point of view. However, your brain has been tricked into thinking it is reality. Lowering your guard by believing in blind-faith is not the way to find the universal truth and reality of science. It is not the path to wisdom. In these boxes, you can focus only on one thing: your religion or a gang's box of beliefs, or philosophy.

If you spend your whole life with the same gang, then you are in a box. Your sense of reality becomes nonscientific. It is distorted because you've learned to see only one perception of the world from one set of beliefs and rules. Under those circumstances, you will not have a vision of the spectrum of life magic. In such a controlling group, it becomes more difficult to see for yourself what is out there. As a result, your beliefs create cognitive dissonance from the illusion. Comparing other points of view is important. You cannot learn from people who agree with you, in an illusion of belief, because they already know that and do not need to be repeated back to you by a preacher.

If your parents believe in illusions and traditions, it does not mean you must believe in fantasy fairytales, because wisdom changes with time. This is your life. You are born a free self-thinker with common-sense searching for understanding to have the confidence of your own free will.

A free self-thinker, in search of reality of science understands that he knows very little in **his mind,** and that reality of science is infinite, because life is meant to be infinitely exciting and challenging; to develop new knowledge, which is what life is meant to be. Paraphrasing President John F. Kennedy about sending a man to the moon, "We do it not because it is easy, but because it is hard, exciting and challenging." This phrase also applies to a better life, not because it is easy, but hard! So, standing up for what you understand in reality is right. Education gives you the tools, but you must be a free self-thinker in search of universal-truth for the time, to be creative, for a better life.

CHAPTER 12

COGNITIVE DISSONANCE

Cognitive dissonance is a blockage in the brain to reality caused by dogmatic beliefs or illusions or propaganda or philosophy given by leaders in their quest for power over others, to keep you idle. When people clinging to strong beliefs are presented with evidence that contradicts them, the evidence will usually be rejected by the "believers." The evidence creates an uncomfortable feeling inside, which is called cognitive dissonance. At that moment, they must protect their firm adherence to their beliefs by rationalizing, justifying, ignoring, or denying anything that does not fit in their belief. For centuries, believers' ignorance enflamed hatred and violence toward other believers from different religions, non-believers.

Throughout history, believers would fight, torture, kill and even be killed to defend and prove their belief is actual reality. That is stupid and it is the aggressive part of cognitive dissonance. It will never be a universal truth, scientific knowledge, or wisdom; that is why the right way is the scientific way to debate and learn from each other, so you can arrive at a more accurate understanding of the reality of science and be more humane today than yesterday. When people are in a state of cognitive dissonance, they are very good at justifying just about anything, even if they are wrong or guilty and they know it! They are unable or unwilling to clear up any obstacles arising from their illusions of knowledge. However, these obstacles have doors that are not even locked, so anyone who wants to can get out into reality, without the need to justify addictive illusion by saying, "All roads lead to Rome." The meaning is that all religions arrive at a belief in God; however, all roads lead to a different illusion of knowledge location. In wisdom, the concept is to analyze the shortest way to a universal truth reality in a scientific method, with no belief or opinion or philosophy allowed. You have to have free self-thinking of individuals searching for original thoughts to understand today better than yesterday.

UNDERSTANDING THE MEANING OF WORDS

Take the word "ambitious." This describes someone who works hard to succeed. The term "greedy" describes someone who cares more about money than anything for more power from greed to compete for an ego and who may cheat, lie, or steal and think of themselves as ambitious. They are not ambitious; they are just greedy, and greedy people are not humane-beings. Words are just a shell that carries a meaning or several meanings, which is why you must ask questions for a more accurate self-understanding of the meaning of words in context. If you don't ask questions, you are left to believe that words with many meanings are manipulated to create propaganda. You must correlate the meaning of words for a scientific meaningful reality in a debate conversation. This leads to a more accurate understanding so that you can arrive at universal truth wisdom to understand your importance.

Most problems today come from people believing leaders who control them by espousing illusions and superstitions over reality. They preach propaganda to believers about what life is supposed to be all about. However, we can no longer continue to be inhumane to humane-beings, or the environment, or other living things, simply because we are unwilling to let go of our many belief systems indoors, by government leaders, or money in return for votes. For example, all people are classified as "human beings," but they do behave like animals and inhumanely. It is time to break the pattern of merely obeying and playing "follow the leader." We are not sheep; we are individuals, and thinkers, and we must each declare our mind's independence to think for ourselves, by the scientific road to universal truths to create wisdom to resolve all our problems for a safer future, because all our issues have been and will be resolved by free self-thinking in God's reality in a scientific method of thinking, **not by being the following sheep.**

Mathematics correlates meaning and facts not words. For example, if the number 2 could also mean 3 or 10 or whatever, it is impossible to arrive at a universal truth unless we search for the meaning that the word two means 2 and the word four means 4, in reality by the accuracy of scientific debate, then you can arrive at a universal truth, for that time, to the correct answer, 2 + 4 = 6. That is an indisputable mathematical understanding, not a knowledge, or a belief, or a philosophy to create a Blind-faith addiction.

All believers of illusions are meaningless. In other words, their justification equals to zero 0 understanding, zero 0 wisdom, zero 0 reality, zero 0 truth, and zero 0 creativity of new knowledge.

Someone has stolen original pictures of my inspiring life, so they can say that I am a liar, justify the illusion of their belief, and attack me back by saying, "Where's your evidence?" That is how terrible cognitive dissonance is. I know who did it. However, I don't judge people, I let them judge themselves.

CONFIDENCE

Free self-thinking in wisdom is thinking for yourself without an organized gang belief of illusion that creates no confidence. Once you understand how to free self-think in scientific debate, you will have the confidence to analyze every period in your life, history, and everything that comes your way. You will be able to see and understand the justifications from cognitive dissonance because you have the confidence from the scientific method of free self-thinking by debating, you can see the spectrum of life reality that gives you the power from confidence to say, "I can fix the problem of homelessness and that is what this world needs now!"

Free self-thinking in wisdom is democratic thinking and America is a democratic country. For example, you cannot swim if you don't know how. **You cannot swim from blind-faith, or opinions, or philosophies, or beliefs, or propaganda.** The reality remains, you don't know how to swim. Ask questions about the realistic way to understand how to swim, once you know how to swim in reality; and by free self-thinking, you will have the confidence for the ability to swim safely.

In a debate conversation, a participant said, "What if after I die, I will go, *whoops*, what I believed is truth?" My response was! "What you said is, that you have no confidence; however, if you free self-think, you will accurately understand in reality that your belief creates no confidence!"

ILLUSIONS

As a free self-thinking lifeguard, I tried to keep people away from rip currents at the beach. Strong outward currents seem calm and very appealing. However, calm water is an illusion of the norm in reality because it will pull a swimmer farther out into deeper waters creating a danger of drowning. So do all illusions of the norm, in reality, creating danger. However, with a more accurate understanding of the spectrum reality of currents at the beach, you will know that calm water could be a rip current that will pull you into danger of drowning!

People must correlate with nature from the reality of science to make everything work and if we don't, we will destroy ourselves.

So, you have no right to complain if you don't do what is suitable from a misunderstanding of life's reality. Because when you are not part of the solution by free self-thinking, you are part of the problem.

Most important among the personal attributes of a free self-thinker is the willingness to accept the possibility of being wrong, which is essential in science to achieve a universal truth, for that time. Once you subscribe to illusory beliefs or this is my opinion or this is my philosophy, you stop free self-thinking and become part of the group dogma. As a free self-thinker, instead of saying, "I believe or this is my opinion or this is my philosophy," **say "I don't know," or "This is what I think."** That means you are still free self-thinking to understand better the universal truth we live in. This leads to enlightenment.

To their people, the Pharaohs of Egypt, ancient Greece, and Rome were believed to be gods or, at least, demi-gods. In Britain, on the bottom of the symbol of English royalty is a motto in French, asserting, that God chose them to be the hierarchy {royalist), and the House of Lords is the house of god's god-players. The fact is that all leaders are neither humane, nor godly, nor chosen by God. They are just people, god-players, intending to have power over people. The concept of democracy and the 10 moral commandments is that nobody has the right to play God or receive something for doing nothing. As people get more understanding of reality, they will want to free self-think, and they will change for a better future.

The free self-thinkers and scientific community tell us the truth about what is happening and warn us about what will happen if we don't change. Scientists should stop preaching and start a conversation with all people to get them involved. However, people are all talked out from learning to preach beliefs, tradition, and propaganda, and memorizing knowledge to listen, believe, and obey. They have no confidence to debate, so they can disagree or even have the confidence to have a conversation about reality.

Fear! makes all animals, or people that are just beings, dangerous. Often free self- thinkers encounter people who are afraid of God and god-players. That is why they join groups of bullies and bullies are so scared and inferior, that they are dangerously aggressive to feel important. It is time to be able to change, and the right of a free self-thinker is important because they do not compete. They don't want to prove they are better; they know it, by understanding the reality of science. That is the reason they always feel good about themselves, live longer, look young and happy, with a smile on their faces, from the joy of a love of life and it is the reason that inferior believers are bullies and hate democratic creative Hebrews!.

All comedians, by exaggerating, are magnifying life. It becomes funny because it creates a more accurate understanding. For example, one day, a mother asked her son, a well-educated young man with the knowledge of words, to watch the front door because she was going next door to a party. The young man watched the door for four hours and got tired of it, so he took the front door off and brought it to his mother at the party. This shows that what is important is the understanding of meaning, not words!

But when you are right, you are right; and when you are wrong, you are in a terrible situation.

CHAPTER 13

FREE SELF-THINKERS CAN RECOGNIZE RIGHT FROM WRONG

In this chapter, I will prove that free self-thinkers in the search of universal-truth can recognize right from wrong; and the ability to belong everywhere and nowhere in particular, because they are searching for universal truth. That is why they have the understanding to create justice.

To prove guilt or innocence, free self-thinkers know that justice must be based on irrefutable empirical and scientific evidence. Unfortunately, today, religious law rules over seventy-five percent of the world's countries. Religious leaders and most judges are positioned to protect hierarchy.

For various reasons, 1) they are elected by hierarchy, 2) they become hierarchy themselves, 3) the accused is judged guilty or innocent according to beliefs or prejudices. Therefore, leaders create many laws, which provide many ways to find the accused guilty. Disagreement with the hierarchy can entail more legal work and that is how lawyers become wealthy—by creating work where none should exist! Understand this.

The really guilty parties are the religious or biased judges who adjudicate based on religion, opinion, or politics to protect their privileges from tradition from believing that they are important.

I saw a conversation on TV between two scholars: one religious believer and one atheist philosopher. They were competing by discussing which was better. While watching them, I realized these speakers were one religious believer and one atheist cross-talking and not listening and learning by respecting each other. Based on their knowledge, the atheist spoke from the philosophy book and the believer spoke from the Bible, instead of listening to what they meant, not what they said. I saw that each party had different conversations because they were competing and not having a debate conversation in wisdom. Conversations within mixed contexts tend to become circular and lose meaning. Just because you are a religious professor, or a scholar does not mean you are open-minded.

OBSTACLES OF THE BRAIN

Doctors and scientists over the centuries naturally thought that the images we see or imagine appear as totally conceived pictures in our brains, like a photograph or a movie. This is wrong. As we now know from scientific evidence, our brain is a digital computer. However, illusions of knowledge and religion distort reality and cause a blockage in the computer brain, like a virus in the body or computer.

Scientists now know that knowledge and pictures create digital electrical signals in the brain, cameras, and computers. Therefore, by creating one's ability to correlate facts into universal truths for the time by accurate understanding and utilizing the power of your mind's free will, you can solve any problem or overcome any addiction because your mind is the real you. Still, your brain is just a computer organ.

Make decisions that you know are right by **humbling** yourself, being willing to learn from others in a scientific debate in reality, and accepting the possibility of being wrong, because nobody knows everything or will ever know everything! Don't let preachers or memorizing religious beliefs that create tradition, propaganda, or knowledge that create illusory obstacles in your brain create misunderstanding in your mind. Aggressive competition proves that they are better hierarchy leaders, who have more knowledge; however, they have no understanding of universal truth for the time, to solve problems.

Today, various treatments can be used to ease or erase beliefs and illusions of knowledge, which create fear, misunderstandings, and violence. Thankfully, these problems acting as obstacles in the brain can now be diminished or removed by using several techniques, such as hypnosis, neurostimulation, exposure therapy, magnetic transcranial targeting, medication, or simply by free self-thinking in a debate conversation to create a more accurate understanding of reality. I prefer free self-thinking in a debate conversation over lasers or medication in my brain!

Your brain is a computer when you search in it. Would you instead find the illusion of knowledge or universal-truth realistic knowledge? For example, when your computer is full of illusionary useless files, they slow down and hinder your work. What do you do? You delete, delete, delete.

A free self-thinker understands that it is wrong for a Babylonian religious leader to ask for money in exchange for a heaven seat. If someone says, follow me for life and the reward will be countless beautiful virgins in heaven, that is not right. It is like saying: "After I kill you, I will give you time to think and talk in heaven." Or "I will make you a saint and give you a great place in heaven or whatever you want after you are dead. They make you believe that they are god, but in reality they are not god!

Why exercise your body? To build strength. *Why exercise your balance?* To improve equilibrium. *Why exercise free self-thinking in search of universal truth?* To understand more accurately and express yourself more intelligently. These things come from your free will.

Let me show you a straightforward way to exercise your mind as a free self-thinker by questioning yourself. For example, *why is it that you think what you believe to be your family tradition is true? Where is the evidence? Why is it that others think their different belief in tradition is the truth? Why is it not possible for all beliefs to be true? Why do believers act as if they know everything? Why can't religious people admit that there are things that nobody knows and that are the absolute universal truth?*

Know yourself, ask questions, seek universal-truth, participate in a debate conversation, and examine everything by a scientific inquiry line. This will help you think, which exercises your mind and facilitates recognizing universal truth. However, philosophy, beliefs, and religions are fixed and stand still, but your mind does not.

Today, many lives are in the hands of warlords, tribal leaders, dictators, or kings; it is not all their fault because some are humane-beings.

Some people say they tried many different jobs but weren't particularly good at anything. However, when they joined the military, they excelled. They were good at yelling and giving orders and thus became officers. What does that tell you? And watch out for that child who is a good liar because he probably will grow up to become a leader in the hierarchy someday, with bloodlines and red-carpet treatment. Just as a gangster must use guns to rob, maim, or kill, a leader must play god be a good manipulator of the meaning, be a good liar, and to give orders. For example, former President Obama made a speech in Egypt and said, "You have the right to your religion." The speech created a fanatic Muslim riot against Christians worldwide, killing many people because all leaders need a religion, making Obama a powerful hierarchy to make war for-profit and more privileges. Leaders' hierarchy will never have the privilege to have the understanding for the wisdom to enjoy the love of all life, which is more joyful than all privileges.

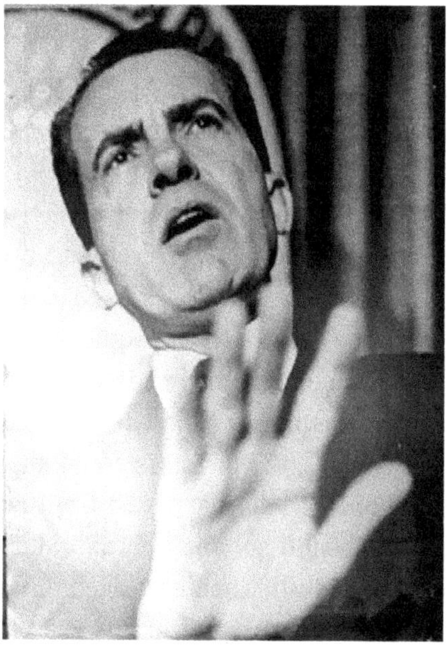

The Voice of Europe took this picture of President Richard Nixon.

Hierarchy makes a pact with religion to gain the support of its obedient followers. Therefore, leaders are evil, not humane-beings. It is the reason that today's modern civilization is so slow to change and adapt to current existential threats! It will eventually destroy us all because politicians who have been in the House of Representatives as Democrats or Republicans for more than four years become part of that gang of believers for their movement and preach half-truths to stay in charge. Former President Richard Nixon was a liar, but not good enough to become a real leader.

CHAPTER 14

VOODOO ZOMBIES

An extreme example of how leaders stay in control and prevent change for the better can be seen in Haiti, with voodoo religious priests who practice zombification, to protect hierarchy privileges.

I am talking about live walking dead people. **This is not a joke!** It would seem to be. In 1981, an article came out in *Harvard Magazine* called "The Secrets of Haiti's Living Dead," written by Gino Del Guercio. Haiti's research by an ethnology-botanist named Wade Davis regarding voodoo and zombies' supposed existence. What Davis found was an underground society of leaders utilizing brainwashing tactics to dominate and control the community. Voodoo priestesses told the people that they would become zombies if they were to go against or disagree with their religious teachings. The community believed them for a good reason. When a person acted against a voodoo leader's laws, the priestess could simply take that person aside and blow a poisonous powder harvested from pufferfish, into his or her face. This neurotoxin would subdue the individual long enough to be presumed dead. The person would then be buried alive. A brief time later, the priestess would exhume that person from its grave. With nerve damage from the poison and brain damage from a lack of oxygen, the person would continue living in the community as a bona fide walking zombie. All this tragic manipulation of men and women is designed to stop free self-thinkers from enabling change for a better future! No country interfered **because all leaders are protected by the United Nations (united hierarchy). Leaders were and are horrible people.**

All religions, in the past and present, utilize underground (gang) societies that use force, torture chambers, and brainwashing lies propaganda. In 1804, Haitian voodoo religious leaders created a revolution that killed and expelled all the white people on Haiti's island to protect their privileges.

So, please don't tell me that religion, makes you moral!

Here is another example of a hierarchy of destructive greed. In Muslim countries, women are for sale to the atheist Babylonian hierarchy that created a religious law. A man. with money can have over 50 women, and 50 men will never experience a woman. However, "nature" makes 103 women for every 100 men. One man cannot satisfy one woman; so, if a man experiences many women, it is okay. However, if a woman experiences many men, she will be stoned to death. Some Muslim people circumcised woman, so they cannot enjoy sex. That is a religious law, not *humane* law.

In 2018, two women from the north of Europe who enjoyed sex went to North Africa to satisfy depraved Muslim men; the women were arrested and killed by Muslim law. Throughout history, leaders have incited followers into killing free self-thinkers. If someone were not obedient to their doctrine of spreading religion, leaders would label the disobedient ones with various names: pagans, heretics, witches, demons, Satan, Lucifer, traitors, communists, and infidels. Leaders created scary books, scary ugly drawings, and pictures of hell for the unbelievers, to show what will happen to you if you don't believe.

Indeed, what exactly was the practical purpose of human sacrifice? Did the blood of disembodied beating hearts guarantee plenty of rain for a good harvest or superpowers on the battlefield? Human sacrifice was a means for those in power to stay in control. Leaders sacrificed many free-self-thinkers who would not follow them. Without courageous free self-thinkers, the community's potential for betterment became less likely. When the ancient Romans converted England to Christianity by war, the English Druids lined hundreds of innocent civilians to sacrifice their old illusory gods, so the Romans would not win. **It didn't work** and religious leaders destroyed all those lives. **Because they can.**

When Spain was going to war with Portugal, the Pope created a peace treaty by giving the northern part of America to Spain and the southern region to Portugal. In return, there was a promise to make everybody Christian, which started the justification for exterminations, that have occurred throughout America and Europe. All today's humane changes and creativity are created by free self-thinkers debating in scientific reality. ...**So, please don't tell me your religion makes you moral.**

The Purpose of Life

When we are born, we have much to learn.

***The purpose of life is: To create free self-thinking to challenge life and by debating, create a more accurate understanding of the spectrum of life's magic reality, to understand God's universal truth of that time, to have the wisdom and the will, to become a humane-being and be able to enjoy life fully,* in a utopia living with humane realistic people!**

PART TWO

You Are What You Do, Not What You Believe You Are!

Always in Friendship

Eleven thousand years ago most people were nomads wandering in small tribes or clans. They came together for safety and to share their experiences and knowledge. This started the practice of hospitality, and by debating in search of understanding, people pondered the thought that some power or force of nature created everything. This was the first universal truth for that time, that united all people, and they named it God. Seven thousand years ago people created competing powerful gangs, the leaders created traditional beliefs, to unite the gang. all gangs' leaders, liars, created power by claiming they spoke with God, to create privileges for themselves. And that started nature's competing animalistic laws, and only competition creates ego, and that created obedient followers, in blind-faith ignorance of universal truth.

The gangs surrounded Israel, their names were the Arameans, the Ammonites, the Moabites, the Edomites, the Philistines, the Phoenicians, and they all claimed they spoke to God. Then came the Babylonian people who created a task to build a tall tower to reach and speak to God. However, they did not find God, and the tall tower collapsed, and everybody died; so, the Babylonians became Atheists to be god players. They decided to write believable books, to explain how everything was created, and said that these Babylonian books were written by God, and made themselves look beautiful to play god, to create more money for more privileges. They became very successful. So, they created the Jews, expanded to create the Christians, and expanded to create the Muslims for more money from violence of words.

Simultaneously, 7 thousand years ago the Egyptians created a powerful empire, the leaders called themselves Pharaoh. Their religion stated that Pharaohs are Gods, and the Sun, the moon, and the stars are God brothers of the Pharaoh, because of their religion, were surrounded by free self-thinkers, astronomers to understand the brothers of the Pharaoh, that become creative; and by scientific debating in God's reality for the search of universal understanding, created many scientific wisdoms in astronomy, navigation, engineering, and built the pyramid, and the best horse chariot of war, and many more! Akhenaton, the Pharaoh, wanted to establish from polytheism to monotheism. Akhenaton understood from scientific research that there is only one God, one Energy, that correlated and encompassed everything in the universe: Akhenaton was killed, because, religion fear change, and religious leaders did not want to lose their privileges, Akhenaton son Moses continue Akhenaton understanding that God was Energy, and created the first democracy his scientific understanding is a universal-truth today.

England wanted to make the Middle East for royalty and peasants. They created the King of Jordan, king of Saudi Arabia. That is the reason Israel was named Palestine, for a King. What they meant was Philistine, it was created by people that don't speak Arabic. Because the letter "P," does not exist in any Arabic words, it did not work, because the world is changing, from polytheism of many god players to monotheism of God's scientific reality in a democracy, for all the people.

So, stop talking about the past because the past was, and is horrible and terrible. It was created by Atheist royalist selfish god players, that created blind-faith addiction.

As quoted, by Mark-Twain, "Religion is opium to the masses,

Today science is changing the world in democracy, by the people, for the creative people that is creating a better world, however, royalty or Babylonian leaders, or religion that created ignorant obedient followers in blind faith addiction that is hard to cure. is escaping, because the dangerous followers are finding it is all a lie and creating their country as a no man land!......

Thank you, Moses for the truth!

- Leon D. Halfon

CHAPTER 15

A STATEMENT OF WISDOM

This is a statement of wisdom. If you believe you are entitled to your exalted position "chosen by God" or feel important, because of your ancestors, you expect something for doing nothing. That is not wisdom; it does not correlate to a universal truth. It is created from illusions of knowledge and a hierarchy to want something for doing nothing by hook or crook, that is a parasite. However, if we free self-think, we will understand that people from all walks of life have generated all knowledge from free self-thinking; you should not be remembered for what you did not do, only for what you did do!

Just believing that you are the right person does not necessarily mean you are humane. It is the actions that you take and the challenges you overcome that define you as an individual, from the action of your free will, which exemplifies your humaneness. Then you will understand that you are a humane person and knowing this reality for yourself will give you self-confidence.

Let me tell you a story: At one time, I lived on 4th Street and 9th Avenue in Los Angeles, California. A neighbor would walk his pet pit bull, and I heard from others that his pet pit bull killed numerous cats and dogs. So, one day, I confronted the pit bull owner and asked, "How do you feel about walking your pit bull?" His response was, "I feel great when I walk this dog. Because I see all the other dogs and cats are afraid of my pit bulldog. They recognize he is dangerous." So, I said, "Next time you walk your dog, put a spiked collar on him and a scary mask on yourself and I will guarantee that not only cats and dogs but everybody else will be scared and you will finally get the help you need for letting your dog kill other animals as fun. So next time you decide to be a follower of a person who enjoys power over other living things, free self-think about it first. Then, discover wisdom to do what is right, and what is right is not to be a follower!"

The entertainment industry could be helpful to help people understand the importance of thinking out of the box. The media can illustrate the positive impact of an accurate understanding to embrace the magic reality of life, which will create wisdom in their entertainment stories; however, the media are in it for money, and leaders are the ones with your money paying to **advertise** propaganda **if people don't free self-think and debate.**

CHAPTER 16

HIPPIES

When new scientific knowledge arrives, we must abandon our previous understanding. That is progress. For example, the "hippies" of the 1960s were free self-thinkers. They were the most significant movement of independent-minded people in the twentieth century. They were not interested in war, money, or power (except "flower power") but instead sought peace, harmony, and love. That is the wisdom that creates utopia.

Hippies understood that control by leaders, gangs, and authorities that created war was wrong; they invested their time in being creative. The status quo community of believers in blind-faith, did not think hippies were right. And yet, it was the hippies who invented desktop computers, created the Internet and the cell-phone's. Hippie's interest influenced Steve Jobs in Eastern mysticism and the underground movements commonplace in the 1970s because the hippie was interested in God reality of science.

The invention of the Internet empowered the individual by instantly opening a world of all knowledge and information at one's fingertips. This made knowledge available without spending a lifetime memorizing knowledge and made some influential leaders and businesses obsolete. Some former hippies helped develop DNA research.

Before the hippies, a highly contagious, deadly disease called smallpox killed millions of people worldwide every year. Until the late 1700s when English doctor Edward Jenner created a successful vaccine that eliminated most smallpox in developed countries. However, smallpox lingered on, killing millions in poor, undeveloped nations for three more centuries until the 1970s, when a former Hebrew hippie, the son of Hebrew immigrants from Eastern Europe, became a doctor in the United States and changed all that. His name is Dr. Larry Brilliant, who, after earning his degree, spent several years studying in India at a Himalayan Hindu Ashram. Back in America, he was chosen to spearhead the World Health Organization's campaign to vaccinate over three million people in Africa, Asia, and South

America, which resulted in finally eradicating that deadly disease from our planet because he was a free self-thinker humane-being.

All hippies were also non-religious. They loved everybody and everything in life. Their motto, "peace and love of life," **is a utopia**. They loved Jesus too, not because he looked like a hippie, but because he was a man of wisdom and a great humane-being. Hippies allowed themselves to be creative and think original thoughts in a friendly debate. They presented opposition to war, governments, leaders, and religions because those forces fear change. President Lyndon Johnson did not think hippies were fellow Americans. Hippies understood we have too many laws that merely serve to prevent the progress of free self-thinking and enrich people in power. This change scares the obedient followers of the illusions of knowledge. Too many scary laws and scary gods are obstacles to free self-thinking.

At the same time, all the leaders, scholars, and followers in the world were saying that it was not possible to eradicate smallpox. **What do you think of those knowledgeable leaders and their followers now?** Today we have the 2020 pandemic, which is just an example! Religious law and hierarchy law created a society of people in this twenty-first century that has the largest population of illusory religious followers and the largest population of very wealthy hierarchy leaders. However, because of all these laws, we don't have a massive movement of independent-minded people.

It seems that we cannot fix anything. The hierarchy is perfect in creating new laws and rules that lock up one million people in jails and the rest of the people locked up at home have a Curfew.

The individual's rights are the most important ones, above any rights of gangs, religions, and leaders in government, because the government must be for the people to create great minds for a safer world!

CHAPTER 17

SATAN, LUCIFER, AND THE DEVIL

The Hebrew people have been writing wisdom books and scrolls for over five thousand years. Seven hundred years after the death of Jesus, Christian scholars' hierarchy collated some ancient "**Hebrew Wisdom Stories"** into two books, the Old Testament, and the New Testament, to create illusionary religion. Meanwhile, hundreds of other books and scrolls were destroyed, including books of Plato, who wrote about Socrates' wisdom and democracy, because these books did not create ignorant obedient followers of religious believers for Babylonian Atheist cruel hierarchy.

Mohammad, who founded the Muslim Babylonian religion, considered himself the "last prophet of the Hebrews," as an allusion of knowledge for the Muslims to believe, whom he called "the people of the books" because they wrote all wisdom stories books of **the Bible.** The Hebrews continued to pursue the search for universal-truth. They are still here because they understand the necessity to love all life and free self-thinkers would prosper, while the illusions of knowledge will be forgotten in the future. Universal truths unite and help people become humane-beings in a Democracy. In the original Hebrew books from which the Old Testament was written, there is no Devil or Hell, only Satan/Lucifer, that were grossly misrepresented by the Babylonian Christian and Babylonian Muslim hierarchy **to** protect privileges! Satan is not a devil. In Hebrew, his name means "someone who disagrees with you in a conversation." To enlighten, Satan's other name is Lucifer in Latin, which means "a bringer of light" in conversation. His names became synonymous with evil and the "fallen angel" Socrates was condemned because he was a Satan/Lucifer. Leaders used the imaged hell with Satan to warn their flock what would happen if anyone disagreed with the religious leaders. Followers who disobeyed were told that they deserved to die and burn in Hell forever, by the Babylonian religion to believe in blind-faith, **and to prevent change.**

A vicious God was a terrifying animal describing God by the hierarchy describing themselves. Hippies and free self-thinkers prefer to morph the word "terrifying" into the word "terrific." Life is not terrifying; it is terrific. Everyone should have Satan/Lucifer's friends in their lives to challenge themselves and let go of biases, beliefs, and fears, that impede their understanding of God's reality of science; it is important to accept the possibility of being wrong when debating with Satan or `Lucifer. However, such conversations are essential to finding a consensus, leading to a more accurate understanding to be better today than yesterday.

The enemy of religious lies is the search for universal truth, to create a more accurate understanding of the reality of life, and that is science!

CHAPTER 18

COMPARING CULTURAL CONTRIBUTIONS OF RELIGIOUS FOLLOWERS AND FREE SELF-THINKERS

Did you know that 66% of all named stars in the night sky come from Arabic, and Arab means mixture of people? A thousand years ago was the Golden Age of the Arabs. During this period, great leaps and bounds were made in mathematics, engineering, agriculture, and seamanship. Star maps were made to improve navigation and mathematics rapidly advanced. The word "algebra" derives from the Arabic "al-Jabr." "Algorithm" is Arabic, and algebra uses Arabic symbols. However, if you look at the Middle East today, you will find many people out of touch with current scientific knowledge and reality. Why is this? A "religious leader" Imam Hamid al-Ghazali (1058-1111 A.D.) was a Babylonian Muslim religious cleric leader and preacher. He is to Islam what St. Augustine is to Christianity.

The Imam created a revised, "proper" interpretation of the Koran—a new guideline for all Muslim people to follow as "religious law **by extreme cruel violence punishment"** for more hierarchy Babylonian power. When this obstacle from religious law happened, Arabs innovation almost completely ended, because of fear from religious law. Innovation by free self-thinking dramatically decreased when Muslim clerics forced people violently by killing the free self-thinkers to create ignorant followers of religious law by Imam Hamid al-Ghazali's doctrine, stating that manipulating numbers and mathematics is the Devil's work. But math is a universal language. Without understanding the language of the universe, how can the reality of science make progress? It cannot. Yet, this happened to Islamic culture a thousand years ago. Today, it is greatly affected by still following religious leaders, who prescribe strict rules of law based on illusions, and by cutting off the head of the creative free self-thinkers who did not follow and obey religious laws to protect Babylonian leaders.

There are approximately 1.5 billion Muslims in the world today. How many of them are among Nobel Prize winners in the sciences? The answer is 0%. One-half of one percent for peace. How many Hebrews are there in the world today? Approximately 10 million. What percentage of all Noble Prize awards in the sciences would you think, free self-thinking Hebrews are responsible for? The answer may surprise you. It is 25%, one-quarter of the world's Nobel Prizes in science. Plus, Hebrews are responsible for 25% of the Kyoto Prizes, 34% of Wolf Foundation awards, and 38% of the National Medals of Science. Medical and scientific advancements have saved an estimated 2.8 billion lives by Hebrews (free self-thinking individuals nomads, in a debate, searching for universal-truth).

Look at the difference between the Muslim religious culture of 1.5 billion Muslims and 10 billion people globally, with only 10 million Hebrews. This means the Hebrew scientific method of free self-thinking is empowered by debating a more accurate understanding of reality. These people who are not followers of leaders of religions, or philosophy, or hierarchy are why the Hebrews have been more than **1000 times more effective in making scientific breakthroughs for a better world. Thank you, Moses, for creating democracy to protect free self-thinkers.**

CHAPTER 19

PUBLIC COFFEE SHOP OF WISDOM

It is time to consider my proposal for a coffee shop named "Public Coffee Shop of Wisdom." To make America great again. It's based on King Solomon's temple of wisdom. It's a public coffee shop dedicated to debating conversations for a better understanding of reality. This coffee shop is a place for all individuals to be heuristic; it encourages people to understand more and learn from each other so they can create more wisdom. A democratic coffee shop that welcomes all people regardless of race, creed, or national origin. A place for social networking, making world friendships and creating free self-thinkers that love all lives from an accurate understanding of reality.

In the coffee shop are books of inspiring scientific literature and history stories of the past, and empty pages in all the books, so readers can write a comment about what they like or dislike about it, no beliefs or opinions allowed. And as with a library, if you want to sit next to someone, you ask permission. You are not to raise your voice, use fighting words or compete, because competition is for animals and preaching creates belief that creates tradition, in blind-faith**. This coffee shop works!** It will unite all the people of this country and the world, for a humane future!

The oldest Hebrew temple in the world is on Djerba's island in the Mediterranean Sea off the coast of Tunisia. Moses himself designed it for free self-thinking and debating conversations to help attain wisdom. Built by individuals, for individuals, but **not for preachers,** it has 3 three names: the House of God, the Temple of Wisdom, and the Greba Temple. Greba is the name of a cake to eat with coffee and debate. The temple's cornerstone came from the original King Solomon's Temple of Wisdom.

CHAPTER 20

PEOPLE OF THE BOOKS

These people of the book understand that there is a creator and they are moral.

In the above photograph, the children of the brother of Moses free self-thinkers humane- Cohen Hebrew people on the desert island of Djerba dress in robes as Jesus did, and are called Jews, because, the story of Hana, is about a free self-thinking Hebrew woman with seven children. She was ordered by cruel Babylonian religious leaders to become religious, or they would amputate her oldest child. As the book goes, they amputated all her children, and Hana did not become religious. It is a very sad book. After that, the people of the book decided to pretend to be religious Jews, so they wouldn't have to be tortured and killed, as all great free self-thinkers of the past. Because life is very precious.

The Dead Sea Scrolls were written 1000 years after the Cohen people escaped to the desert island of Djerba, because the Cohen people

did not want to be religious. They don't dress alike and are not hugging each other because they are not part of a gang. They created the Kosher method, chucking the inside of the animal, and if the animal is not healthy, it is not Kosher, and is sold to the Muslim, that started people wanting to know what they are eating. They were the first doctors, because they are free self-thinkers, individuals in search of wisdom for a better world. The first famous Hebrew doctor was A Rambam, he wrote "70 Rules to Stay Healthy." These rules are still good advice to stay healthy today. You cannot tell who represents the people because they are all individual free self-thinkers in a democratic society that doesn't compete**. They are selected by the tribe of Levi, the direct descendants of Moses as the most humane**. If you strike them, they will turn the other cheek, just as Jesus taught, because violent people are animals, not humane-beings! Unfortunately. We are still cruel to living things today and many people blame God for it. However, these teachers of wisdom, the Hebrew Cohens, are examples of humaneness. They've lived on this island for 2,500 years, didn't overpopulate, had no weapons, wars, jails, lawyers, or judges, or any violence; and disputes would be settled by a wise man creating a more accurate understanding by a scientific realistic debate for the wisdom of universal truth for the time, and that is utopia. **The followers of Abraham are Hebrews, because word correlate, and the 3 faders of the Hebrews, are Abraham, Isaac, and Jacob. Today, the Muslim leaders are screaming at the Jews for not being Jews, and for being a democracy in Moses' law, searching for universal-truth in science, because today 65% of Israelis are not religious Jews. As more of the world is becoming free self-thinkers, it is a threat to the religious leaders; and that is the reason they are preaching hate of Israel.** The "People of the Books." Their ancestors wrote the books of the original Bibles, the books of the New Testament, the Old Testament, and many other books of wisdom. Today, the Bible is found in hotels, motels, and homes around the world. Over time, the Bible was used to teach people how to read and write. These Hebrew Cohen people worked as masons, carpenters, and jewelers because they **did not want anything for doing nothing**. They have the wisdom to be humane with each other. They do not value the power of money, hierarchy, or competition that creates the ego. When the ancient Romans ruled, they grew wheat for their Army in North Africa and lived on my native desert island of Djerba. At that time, the Romans had conquered most of Europe, North Africa, and the Near East, including present-day Israel and King Solomon's Temple. That is where Jesus kicked over the "money table of the Atheist Babylonian religious hierarchy rabbis of the Jews" because the Atheist Babylonian Rabbis were greedy for Roman money, by competing for an ego. That is the reason the Hebrew Cohen people escaped to Djerba.

John the Baptist and Jesus told universal-truth wisdom and for that, they did not deserve to be killed. Unfortunately, Atheist Babylonian leaders do not willingly give up their power of money and privilege, no matter who, or how many, or how much they destroy. Do not blame anybody else but the hierarchy in charge, because the people are just free self-thinking humane-beings and wore crucifixes. By crucifying the man Jesus, the Babylonian Christian rulers, turned Jesus into a magical god, and in effect, set up the hierarchy of the early Atheist Babylonian magician Christian church that Jesus disapproved of. This cruel religious hierarchy runs most of the world today, using propaganda and enforcing religious cruel laws. By the time people begin thinking for themselves, it is often too late. The hierarchy becomes self-propelling and is maintained by their blind-faith ignorant followers who believe in aggressive inhumane obedience and the power of money. However, the Cohen tribe survived because of their wisdom by pretending to be religiously tolerated by world power's hierarchy.

My grandfather and uncle were the representatives, pretending to be Rabbis of wisdom for the community on Djerba's island during the 20th century. Since Djerba was unknown, no hierarchy came to destroy the ancient books of wisdom in the extensive library that my grandfather owned to control the world. All the books were handwritten and rewritten because these originals were thousands of years old. From these books, I learned all I could, about the great wise inspiring men of ancient history: Moses, King Solomon, Cyrus King of Persia, John the Baptist, Jesus, and the Knights Templar. All these free self-thinkers understood that "you are what you do!"

My grandfather's name was Simon Cohen. He created the Glatt Kosher, which admonished people to treat animals humanely. If an animal is killed for food but dies in pain, then it is not fit for consumption because it suffered. To be considered a Glatt Kosher food, the animal must be slaughtered painlessly with kindness and respect. Initially, the Jewish leaders did not accept the Glatt Kosher killing method because it was not economical. However, more people are humane today and often demand that their food be Glatt Kosher. They want a healthier mind and body because **it creates healthier meat, with no adrenaline,** and to be humane toward animals all their lives or become vegetarians, as, a grand-San of Simon Cohen, I demand that Glatt Kosher be organic, because, when Simon Cohen created the Glatt Kosher method everything was 0rganic.

The Cohen people paid a bribe to the Muslim hierarchy and paid sixty-five million dollars in gold to the Babylonian Germans to be left in peace!

After World War II, the Hebrew Cohen tribe departed Djerba to Israel. They wanted independence from the cruel Babylonian Muslim hierarchy and their followers in blind-faith and the fanatics inferior gangs; that is the reason… they are bullies, from the need to be important and attacked innocent people. These murderous thugs bombed the Greba temple, killing fifteen people, in blind-faith.

So please don't tell me that religion makes people moral.

ANCIENT BOOKS OF WISDOM

I was very fortunate to read historical stories in books of wisdom early in my life, and then debate them with others in the Temple of Wisdom. To understand knowledgeable meaningful conversations with my grandfather the wisdom **of great free sell-thinkers of the past**: All these books of the New and Old Testament and other books of the Bible are historical. To understand these writings' meaning and wisdom, they must be analyzed in a debate conversation in a scientific reality to find the original meaning. The best way to find out the meaning of words is always to ask questions about their meaning. For example, when was the book written? What was it like when it was written? Why was it written? What did the writer mean?

CHAPTER 21

COMPETITION

Ever since I was young, I have never enjoyed competing with other people because of my accurate understanding of reality. I concluded that competition enlarges the ego of the winner and creates inferiority for the loser. On the other hand, I have constantly challenged myself to become better and better at everything I do. For me, this understanding helped me grow humanely. It may seem that competition brings out the best in people to be useful for hierarchy. However, if you free-self-think in wisdom, you realize the universal truth is that the greatest humane-beings in history did not compete against others. Instead, they challenged themselves to be better and received boundless joy from the inspiring journey of life with a more accurate understanding of what is right from God's reality of science.

That is what makes the world a better place; for example**, Jack Lalanne did not compete, he challenged himself to become better and better at everything he did; and helped everybody free self-think, to challenge themselves to be healthier. He did it from his love of life as a humane-being. I don't care how big of a competing champion you are, you will not be as powerful, and help as many people as Jack Lalanne, because you are selfish from competing ego.**

Another example is, **My friend Henry Lamothe. His performance was to jump from 35 Feet high into a pool of water 14 Inches deep, and 10 Feet wide. He said that if they introduce this sport in the Olympic, it will get rid of all the egotistic competing champions. You are what you do for the love of life, not from believing you are a winner, because most people are brainwashed by leaders that the word I, is ego. However, ego derives from only winning the competition. It does not derive from the word I, or from what you did, or what you do.**

Our competitive behavior is the reason we are here; it comes from nature's animalistic law. Nature's law is cruel and arbitrary for humane-beings, but it is not too cruel for inhumane-beings who act like animals. People are corrupted by aggression, which derives from nature's competition. It seeks only to dominate and survive. It is a result of this awareness that, over time, people have made strides by free self-thinking in search of universal-truth to become more humane, for a creative safer world.

In New Guinea, tribes conduct annual war "competitions" against each other's tribes. If tribe A kills more warriors than tribe B, then tribe A wins. The chief says, "It's just like a football game, it is tradition." Then losers will feel inferior and inferior people become bullies to feel superior. In nature's law, the loser ends up being eaten by predators. In wisdom, you don't need to compete or prove anything to anybody but yourself.

Free self-thinking in wisdom enables us to understand that competition, religion, tradition, gangs, and greed do not correlate with the environment or democracy or God's reality of science. We know this is true because the homo sapiens hierarchy has destroyed every great civilization in history by wars and environmental degradation. Unfortunately, by competing and polluting, people are still doing that! Nature needs humane-beings to protect all life and the environment. So, for example, if someone is humane, they would raise a dog to be compassionate. But, unfortunately, not all dogs are raised humanely because they are bred to compete, fight, and kill other dogs.

One of Moses God's Commandments is "Thou shalt not kill," but that humane law hasn't stopped people for thousands of years from killing by believing in blind-faith. Per my grandfather, free self-thinking is a scientific method for a more accurate understanding, which leads to wisdom. Many people have persecuted others throughout history, believing they were human, and others were not human. However, persecutors were the ones who were not humane-beings, and the word human is created by religions that make you believe you are important. You must give up all your paradigms to advance to the next level of attaining new realistic understanding by free self-thinking and debating to become a thinker. Not a stinker! A free self-thinking humane-being would never be proud of hurting anyone, even in self-defense. You should only be proud of the lives and the environment you nurture. Ignorance of reality is wrong for people, no matter how you justify it. Because it is proven not to work, in reality of science.

CHAPTER 22

TO BE A HUMANE-BEING OR NOT TO BE A HUMANE-BEING - THAT IS THE QUESTION?

Not all people are humane-beings. However, all people are borne free self-thinkers with common-sense, and some people are encouraged, by the adults, to search for universal truth in reality, and by asking **QUESTIONS**, created with the accurate understanding to love everything in life, and evolve into a creative humane-being to create wisdom, which makes all knowledge, for all lives. This scientific system in God's reality is proven to work. However, the majority of the people that are born free self-thinkers with common-sense, are obligated by the adults to believe religion's tradition to unite the family to make adults important or make baby stay idle, as mama's baby. That creates inhumane people who torture and kill free self-thinkers and humane people who are making and will make a better world.

Regardless of the justification for fighting and killing, it is **stupid**. Today, we have the United Nations to protect countries (hierarchy) because hierarchy created it. However, we need a global computerized high-court program run by humane-beings to protect all individuals' lives and the environment from the United hierarchy. To become a world **of humane-beings or not to become humane-beings, that is the question?**

It doesn't matter where you come from, your background, or your skin color. You can become a humane-being. The only option is to bring people together in a public coffee shop of wisdom. by using debate conversations in a scientific method of reality to find the universal truths for that time to understand life's energy, the Creator, which will make all people know right from wrong, and become moral humane-beings.

CHOOSE CONVERSATION IN REALITY INSTEAD OF VIOLENCE

As an animal trainer, I understand that the wilder the animal is the more the animal behaves by aggressive instinct, and the more they misunderstand you, the more they become frightened and more dangerous. This is nature's survival technique.

I grew up with the Hebrew People of the Book on the desert island of Djerba; they experienced no violence among themselves for thousands of years. When people understand the importance of a scientific method of debating to find wisdom to become a humane-being, it will enlighten the rest of the world to solve all problems in the reality of science. In real estate, it is all about location, location, location. The Hebrew People of the Book decided to separate themselves from the rest of the world that lived by natural law, which naturally creates a balance of overpopulation by violent competition of wars. The people of the Book chose to live in the isolated location of the desert island of Djerba. They made a peaceful existence (utopian) by utilizing debate conversations to discover more wisdom from universal truth and are an example for the rest of the world!

About 2 million years ago, people and chimpanzees had the same ancestors. One group of ancestors made up words for conversation and evolved to take over the world. Unfortunately, the other primates' group was unable or didn't want to communicate and remained a gang of chimpanzees still living in nature's violent law.

Today, people are divided into two camps: The believers' followers of illusion hierarchy gangs that handle life with natural law, as chimpanzees do; and others as humane-beings, solving their problems with scientific minds by debating conversations.

One solution is to separate the two groups with walls, people in illusion on one side and humane-beings on the other side, to avoid inferior bullies' confrontation or create a separate part of the city as a Ghetto, to create a safer location for humane-beings and that will eventually create a revolution or riots for more demand from bullies by aggressive cruel leaders.

CHAPTER 23

DAAA

Because we are homo sapiens, we can use many words to talk; other animals cannot. So, when we have meaningful scientific debate conversations searching for universal truth, we evolve into humane-beings, not just animals that can speak. We are more advanced than animals because we can use thousands of words to understand a conversation's meaning. But even so, people often misinterpret the meaning of words. In the future, our languages will evolve to give every word only one definition, one specific meaning, like mathematics, for a more accurate understanding of meaning in a conversation. If someone attacks you in a conversation, it is because they don't understand what you mean. They get this feeling that you are talking over their heads. For example, suppose two participants are talking about something you do not understand in a conversation. In that case, you might ask them to stop because it creates zero thinking, zero learning, and zero understanding. If you want to learn, it is up to you to free self- think and ask questions to develop a more accurate understanding.

In today's world, dictionaries and leaders give many meanings to a single word. This creates misunderstandings of reality in a propaganda program to make people uncertain and that creates people with no confidence, for leaders to have more power over their followers. Leaders love personal profit at the cost of their deluded believers in illusions of knowledge and opinion. Dictionaries and religious books are complicated so you will need help from a leader or a preacher.

However, the true meaning of words should be simple in knowledgeable debate for an accurate way to understand. That is why I create new words to give more precise meaning and interpretation. I introduced unfamiliar words in this book that you won't find anywhere else to clarify this book's intentions. For instance, my books are the only books that are using the word "Daaa."

On the other hand, people who establish their thinking for themselves and get their independence call themselves free self-thinkers. Throughout history, people moved to new locations to establish themselves and then declared their independence. For example, the United States declared its independence from England. Then Texas, California, and New Mexico declared their independence from Spain. Israel declared its independence from Jordan; also, part of Scotland wants independence from England because they understand the illusion of divine royal leaders and the Scots do not want any part of it. They recognize that independence should be available to all people. However, some people don't want their independence because they are insecure from listening to scary preaching, and propaganda so they need to be cared for, in exchange for a vote.

People who did not get their independence and are part of a collective influenced by religion, tradition, blind-faith, propaganda, a dominating family, or a gang, are called "Daaa." The television show "Star Trek Next Generation," created the word "Borg," because they are boring people in a box, or by believing in a square book, part of a collective by tradition. They are created by parasite leaders in power that need Daaa and despise free self-thinkers and the countries they live in. Rulers hate the fact that free self-thinkers cannot be controlled and influence the people that are Daaa. To maintain power, leaders must destroy free self-thinkers by creating hateful propaganda, to protect their privileges. In the mid-20th century, Stalin's communists (atheists) killed over 50 million free self-thinking people in Russia to stay in power. Stalin said, "What is important for me is my family (his gang of henchmen)." He was not a representative of his people. He was a murderous tyrant. Hitler, the mid-20th-century German Fuhrer, had his followers kill over 30 million people to stay in power.

Similarly, Turkish rulers killed Greeks, Muslims, Kurds, and Armenians to maintain power. They did it with the help of ignorant competing mobs of followers who were Daaa. China's huge population was peaceful until the Chinese communists (atheists), led by Chairperson Mao Tso Tung, killed over 70 million of his people. The average Japanese individual before World War II was not warmongering. Yet Japanese leaders ordered the murders of over twelve million Chinese and Filipinos. In Africa, the government of Rwanda conducted massive genocides as well.

Today, there are too many free self-thinkers to kill them all because of the American Constitution's wisdom: By the people, for the people from the people. Not for hierarchy, or religion, or tradition. However, we need a lot freer self-thinkers in wisdom to stay safe from leaders, and their Daaa.

Peace-loving religious people have been made irrelevant by their silence because they are not free self-thinkers. They will be awakened one day in the future and find out that the ignorant Babylonian leaders have taken over their world and their nightmare will begin! Peace-loving religious Syrians, Koreans, Germans, Japanese, Chinese, Kurds, Russians, Rwandans, Serbs, Afghans, Iraqis, Philistines, Somalis, Nigerians, Algerians, Lebanese, Jews, and many others, have been killed because they did not speak up until it was too late! They followed leaders blindly instead of thinking for themselves in a scientific debate to know right from wrong, and that is still happening because they are Daaa.

When the ancient Greeks were contemplating going to war against the Persians, the people wanted to know from the religion Oracle at Delphi who would win. The Oracle's prediction was, "If Greeks go to war, you will destroy a big country." With this prediction in mind, the Greeks confidently went to war and were defeated. The Greek people complained that the prophecy was wrong. The Oracle (religion) responded, "You had a big country, you fought a war, and you destroyed your big country." The prediction was correct; the Greeks just misunderstood the message because they could not ask the Oracle, or preachers, or lieders, any questions or debate which big country, they were talking about. When in doubt, always ask questions to understand the meaning.

VERIFIABLE CONCLUSIONS

Another phrase I like to use is "verifiable conclusion." In a scientific method, a verifiable conclusion becomes a universal truth for the time. This represents a scientifically proven theory. It is not blind-faith, a belief, an opinion, or an illusion. As an example, Darwin's Theory of Evolution is not a belief; it is a verifiable conclusion about all lives including people evolving differently to adapt to nature or to become extinct**, because nature doesn't adapt to life, only God the energy of the universe can create life energy, and life is real, not an illusion. It is a reality.**

Nature methodology is what nature does with everything. It turns mountains into valleys, dead trees into wood dust, and dead animals and people into the dirt. Nature's methodology breaks down objects, it does not and can't create life. **Only** life energy can create life, **only** God's reality of science evolves to create a more realistic understanding of life; and evolution only creates changes in all species to adapt to the environment. For example, nature can't create an intelligent computer, it can **only** create dust out of the intelligent computer. **Only** intelligent, of higher consciousness, by free self-thinkers, and debating in the reality of science can create intelligent computers, because there are cracks in the energy of the universe that create enlightenment of scientific debate, for the light in your mind to create new knowledge from an accurate understanding of God's reality of science.

Why must anyone be prejudiced against others who have a different skin color? Do you know what causes various skin tones? The cause is vitamin D. For example. If you live in a sunny area, your skin becomes darker to protect you from too much vitamin D. If you live in an area that is not sunny, your skin becomes lighter to absorb more vitamin D. Similarly, my horse grows longer hair in the winter to protect him from the cold. Shorter hair in the summer to protect himself from the heat. Also, if you don't get enough or too much vitamin D, you will get sick, because nature doesn't adapt to life. I am telling you this because it is wrong to be prejudiced against people just because their ancestors had different levels of the sunray or having a different point of view of the spectrum reality of the life magic and by not being hierarchy and by not competing, you will create a more accurate understanding of life reality. Some believers believe that Darwin was wrong, because of the cognitive dissonance in their brains from believing in illusion and refusing to accept God's science's reality. However, people that memorize science to create a religion from science cannot understand how you can exist without believing, from competing ego. So, they justify their illusion, and justifying is not an answer, it is just an excuse!

CHAPTER 24

THE UNDERSTANDING MOVEMENT

If you want to get elected in America; you must seek religious leaders' support because they can influence their followers' votes. This is tremendous power for preachers who tell their flock what to think and do. This is influencing people, and it is wrong. In a real democracy, people are supposed to free self-think and debate to create understanding! Not follow and obey! Ironically, every living American president, because they are a hierarchy that needs religion to stay in power, which is the only reason they attended televangelist Billy Graham's funeral because, he could make 2,000,000 people vote one way, except for Donald Trump, who did not participate in the funeral!

Abraham, Moses, King Solomon, John the Baptist, Jesus, Buddha, Socrates, Archimedes, Newton, Galileo, Einstein, Martin Luther King Jr., Rosa Parks, Steve Jobs, and many more were free self-thinking humane-beings of different eras, various places, and various skin colors. They did not align themselves with religious leaders and hierarchy to become more powerful or improve their lot. They were not followers or fundamentalists. Instead, they opposed religious autocracy and helped humanity adapt.

In today's world, cruel religious armies are not needed. That is the government's job that represents all the people's inhumane laws and fosters all the people's movement toward fixing all problems in the wisdom that works, with instant access to information about our interconnected world, and to accurately understand God's scientific reality.

The word "ignorant" derives from the word "ignoring scientific reality." It is not meant to be an insult. It simply means you are ignoring God's reality of science, which is out there. The term "knowledge" comes from "knowing," not from the word "believing or opinion." Because if it is only a belief or opinion, then it means you do not know. Bibles say God created man in His image. What the Bible writer meant is that God is energy. Man is energy because what is essential is meaning, not words.

CHAPTER 25

THE ONLY BIBLE STORY ABOUT SATAN IS IN THE BOOK OF JOB

What is the real meaning of the Book of Job in the Bible? When was it written? Why was it written? What did the writer mean? Ask yourself and in a debate, these questions to learn from others will create a more accurate understanding of meaning. For example, when the story was written, pharaohs and kings were believed to be gods. But the god of universal cosmic energy is believed to be the God of all Gods.

The man Job was a good and prosperous father, living in Egypt thousands of years ago. He had property, animals, and children. He prayed ten times a day to show his love of the Pharoah "god-king" and praised him for his wealth. In those days, Pharaohs had advisers. One adviser was a Satan, meaning in Hebrew, someone who disagrees in a conversation. To prove his power; the Pharoah egotistically boasted to his advisors that Job would love him, no matter what. Satan disagreed with god-king and said that Job would stop praying and loving you if you took all his wealth.

Despite this advice and to make a point of his ego, the Pharoah god ruined Job by killing his three daughters and seven sons and confiscating his property, animals, and valuables. Afterward, at first, Job continued praying to the Pharoah god-king, saying philosophically, "The Lord gives and the Lord taketh away." Then Satan pointed out to the Pharoah that Job was not physically hurt and advised him to stop loving you if he hurts Job or sickens him. That advice challenged the Pharoah's ego, so the god-king proceeded to violently inflict painful boils and open sores all over Job's body. His neighbors and friends believed that Job deserved this punishment because he didn't pray enough to show his love for the Pharoah god-king. Job reminded his neighbors that he still prayed ten times daily to the god-king.

Later, while Job was alone and full of pain, he supplicated to the "God of all Gods," the energy of the cosmos. "How could you let this happen to me? I kept praying to the Pharoah religiously every day." The cosmic response was, "I am the life energy of the universe that made a Goldilocks' atmosphere for life to flourish and gave you higher consciousness, for you to appreciate all the beauty of living things; and instead of being grateful, you expect me to listen to your prayers and complaints.

Life is not about what I can do for you. It is what you can do for life. Kings and religious leaders (god-players) who control others need their ego to be worshiped and pray to feel good about themselves. However, I am the life energy; I do not want or need to be praised or worshiped or prayed to. I gave you a higher consciousness to understand, enjoy, and have the wisdom to take the responsibility of protecting the Goldilocks' life for your own good. Do not blame the universe's life energy because all the problems are created by people believing god-players. For example, all religions believe in a God in the sky that is a man or a woman because it was created by a man or a woman (god-player).

In the city of Jerusalem, there is a place called One Hundred Doors. Behind each door is a small shop and, in each shop, a woman is selling small items to make money to feed her husband, who does not work or do anything profitable. He prays every day to keep God from destroying the planet, but he is deluded by religious beliefs and their illusions of knowledge. This is not wisdom. It is a man's religious law to control and abuse women!

A Hindu friend of mine from India named Gar, who I met while doing a construction job, was having domestic issues, so I invited him to live in my house in Los Angeles while I went to Texas. Gar watched over my house, and we talked on the phone about everything. He would tell people, "Leon is my guru." Likewise, I thank all my friends for being my gurus (my Satan). It is all about the importance of knowledgeable, meaningful realistic debate conversation. You must give up all your paradigm that makes you believe you are important! to advance to the next level of attaining new realistic understanding by free self-thinking, to become a thinker, not a stinker.

For example, my friend John Niemans, in a debate searching for universal -truth, is my Satan. If you are searching for your cat, do you tell the accurate truth of what the cat looks like, or do you tell what you believe the cat looks like? That is why People in search of universal-**truth,** in a scientific debate, are the only one that will always tell you the **truth**!

CHAPTER 26

DISCOVER IF SOMEONE IS A FREE SELF-THINKER SEARCHING FOR UNIVERSAL TRUTH

A free self-thinker searching for universal truth will respond in conversation by telling you what they understand from their conclusion; and, if they don't understand to prove their conclusion, they will not say this is my opinion, or I believe or lie or compete, because it is against the search of universal-truth. A believer will respond by telling you what someone else has spoken, written, or believed. For example, Jehovah's Witnesses will not tell you what they think or understand; they want you to read the Bible and believe in blind-faith, or if you want to save the world from parasite hierarchy, you have to be anointed with oil by a religious hierarchy to be the savior. Or if you want to be the president of the country, you need the blessing of a religious hierarchy. How convenient it is for parasite hierarchy to stay in charge by keeping followers ignorant of truth.

PEOPLE WITH LITTLE OR NO FORMAL EDUCATION

Richard Branson—One of the most recognizable billionaires today, struggled with dyslexia and dropped out of school when he was 16. He founded his first business, a mail-order retailer called *Student Magazine* company would become Virgin Records and its retail stores. During his career, Branson ran approximately 500 companies.

Thomas Edison—America's most prolific inventor, did not graduate high school. Instead, Edison's family moved to Port Huron, Michigan, where he attended public school for 12 weeks. He was a hyperactive child, prone to distraction, and he was deemed "difficult" by his teacher.

His mother quickly pulled him from school and in a realistic, debate conversation taught him at home, whereby he learned scientific debating in knowledge conversation. Thomas Edison became a genius.

David H. Murdock –After dropping out of school in the ninth grade he worked at a gas station until he was drafted into the Army in 1944. After returning from World War II, Murdock purchased his first business, a diner in Detroit. He sold the restaurant several months later and made a $700 profit. In 1985, he acquired the Hawaiian real estate outfit Castle & Cooke, which owned the fruit company Dole. As a result, Dole became the world's largest fruit and vegetable producer.

Steve Jobs- His family escape Lebanon to a democratic, U.S.A. He dropped out of college after attending for a short time. He started and developed Apple computer products.

Robert Riley was born an Irish Catholic **and did not graduate high school. Instead, he wrote articles about new designs of cars and boats for "Popular-Mechanic" books for 30 years. His last car design did 225 miles per one gallon in a world race of the most economical car of all cars**. The result was, in search of universal truth reality, he became a genius, not from illusionary religion!

Orville and Wilbur Wright—Neither one graduated high school but they started building bicycles. Then they decided to build a motorized glider. They designed the first propeller for a flight. By debating for 4 years, they built real airplanes, not gliders, and founded the first airplane manufacturing company in the world; it was named Curtis-Wright Airplane Manufacturing.

Albert Einstein left school at the age of 15 years old and received his doctorate degree, at the age of 60 years old, with no education for it.

Quentin Tarantino was a Movie Director, with little formal education.

Vidal Sassoon was a Famous Hairdresser, with little formal education.

Other famous people with little formal education who created wisdom by free self-thinking are **Benjamin Franklin, John D. Rockefeller, Walt Disney, Charles Dickens, David Karp, and many more. All these people were into scientific, accurate understanding, in search of reality, to become creative from their wisdom, not scholars or by memorizing knowledge.**

It seems that the most creative people are not scholars because knowledge cannot create new knowledge! On the contrary, they are free self-thinkers searching for universal truth in scientific understanding, in the spectrum of God's magic reality of science to create new knowledge! So please don't compete with your knowledge because your knowledge can't come from illusory religion or propaganda and tradition, to create no confidence, fear, jealousy, hate and separate people, because you are looking at the magic of life from one direction, which creates obstacles in the brain to promote parasite hierarchy with no understanding or scientific reality. Only by free self-thinking and debating to understand from others the spectrum magic of life reality, you will arrive at the universal truth for the time, to be creative, and unite all the people.'

Two friends are in a coffee shop in America; one friend tells the other, "My son memorizes knowledge like a computer; now by competing, he became a senator. By more competing, he can become the president." The second friend said, "My son debates in a scientific method of thinking to understand the spectrum reality of life. He became a humane-being. And one humane-being called Jesus was selected as a God by the scholar hierarchy because of his understanding wisdom!

People believe that affiliation with their gang, tribe, religion, tradition, or philosophy, gives them an identity, but it is only an illusory one. By free self-thinking in search of universal truth, you will not lose your identity. On the contrary, it gives you a truthful realistic personal identity, because you are only what you do. When you do nothing, in reality, you are nothing. Only people who are creative by understanding God's reality of science, make a better world, and they deserve it all! Because this is what we all need today, creative free self-thinking people for a safe world. **Why don't you get over your blind-faith illusion addiction and start searching for universal truth, and that is, the most important.**

When children are repeatedly told something at an early age, they believe it. Therefore, some people who have been encouraged become athletes, artists, scientists, and other vocations. If you are told by your elders to believe that some people are crooks, you will grow up to believe that they are crooks, or sinners, or inhuman. Only with your free will and determination you will understand how to achieve goals to overcome challenges.

Since the beginning of recorded history, people have gathered together to avoid predators and other large groups. They knew it was wiser to be part of a collective for security. Nowadays, it is wiser to avoid big groups and stay safer as part of a smaller group of people you know, enjoy, and trust in scientific debating rather than being in a mob with a leader! It is time to change into a new universal-truth for these times,

To make a better America, we should encourage free self-thinking for all people to "get out of the box" of their restricted precepts. This will adjust the way people see things and help them make the right decisions. We should all be grateful to live in a country that allows us to experience and enjoy the full spectrum of the magic of life and live among creative, free self- thinkers as an example to the rest of the world.

People must live by humane law and the highest achievement in life is to be a humane-being. Wisdom and universal truths are attainable by your love of all lives as a humane-being, learning from each other without competing. Competition is an old pattern in nature's laws from the past when people acted more like animals and still do because they don't know how to learn from each other! **For example, in a debate, you must treat each other with respect, so you can learn something brilliant.**

CHAPTER 27

DISCOVER A NEW MEANING AND A NEW WORD

If you discover a new species or find something new and there is no word for it, you give it a name. That's how all words are created. Explain the meaning and it is understood. Only then, meaningful conversation and clearer communications are possible. However, suppose you are reading law documents, books, or any writing that you don't understand and there is no explanation for that word. In that case, dictionaries and research may be necessary if you wish to be more accurate in reality, during a debate conversation; so, keep on researching and asking questions of meaning.

If the Bible is the greatest story ever told, then Babylon's story is the greatest example of a universal truth! The story is about a group of Babylonian people who created a task to build a tall tower to reach to speak with God. The God in this bible tale did not want living people invading his heaven. When the Babylonian builders completed their mighty task, God made each of them speak in different languages, causing misunderstanding. The workers began babbling incoherently and failure to communicate caused the building to tumble down, killing them; this kept them from doing their tasks. The wisdom here is if you don't communicate clearly and understand each other's accurate meaning from believing in religion and tradition, we will never arrive at an understanding of God's universal truth of science, and we will **never succeed at a better world, for all lives!**

Believing this religious story in the Bible, as fact, is ridiculous. It happens because Babylon's leaders have knowledge and think they know everything. They will never arrive at an understanding of God's universal truth of science, and they will **never succeed. That is the reason they are Atheist leaders of lazy people; however,** if you look at it as a free self-thinker, you realize **it is the greatest wisdom story ever told. Because by changing books for** Babylon's **hierarchy greed, and without communication and accurate understanding, we will never succeed in our task!**

If you accept from your illusion of knowledge that God punished people trying to reach His heaven, this information will create obstacles in your brain.

Here is what has happened, and what will happen in Europe, and is already happening, all from Babylonian leaders' greed.

Lebanon was Paris of the Middle East and wanted to be an independent democracy. However, democracy does not need leaders, and the leaders of Lebanon did not want to lose their privileges. So, they voted in the United Nations for democratic Israel to get out of Lebanon, and that created a massacre of the democratic Lebanese French-speaking people because the Lebanese leaders did not want to lose their privileges. Eventually, the Lebanese leaders were killed and replaced by Babylonian Muslim leaders. All leaders would rather die than become a commoner or a person from the people, of the people, for the people. Steve Jobs' family escaped Lebanon to America. He started and developed Apple computer products. He was a free self-thinker of Lebanese descent. New knowledge is not created by religious belief in blind-faith addiction, or illusions, or propaganda, or leaders, or philosophy, or knowledge! New knowledge is only created by searching for universal-truth in a democratic country.

The goal of the Muslim religion is to convert all the people of the world into the Muslim religion, no matter how many they kill for atheist Babylonian cruel leaders. So, watch out. European people from being converted by cruelty into the Muslim religion by people that are in blind-faith adduction.

> **"Learn from yesterday, live for today, hope for tomorrow. The important thing is not to stop questioning."**
> **-Albert Einstein.**

POLICE ARMOR SHIELD

The first police armor shield had the symbol of a six-point star on it. In search of wisdom, the symbol originated when the Hebrew People of the Book wanted to find a symbol of God to put on David's shield to protect him from the Philistine warrior Goliath**. The Cohen Hebrews, people of the book, understood from Moses that God was Energy.** As they looked at the sun shining on a bush, it created a six-point energy spark. They understood that this was the fingerprint of life energy God. What we know today from science is that most snowflakes are six-point. When you put five-point snowflakes on a Christmas tree, you become ignorant of reality creating blockage in your brain, because Five-point snowflakes don't exist in reality. All building blocks of living things are SIX POINT STARS.

This is not a religious belief; it is a realistic scientific method of searching for universal truth by free self-thinking Hebrews. That is why they put that star symbol on David's shield to help protect David from Goliath by the fingerprint of life energy, and that is not a Jewish star, or a David star**, it is the fingerprint of life energy God.**

The first police badge was a six-point star that the sheriffs wore. When officers wear their badge, it represents God's symbol, the Energy of the Universe, to create humane laws of wisdom to protect and serve the innocent free self-thinkers from aggressive, violent, inhumane inferior bullies!'

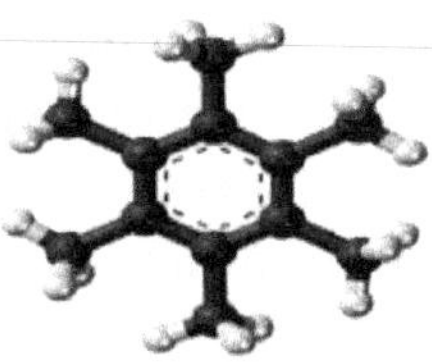

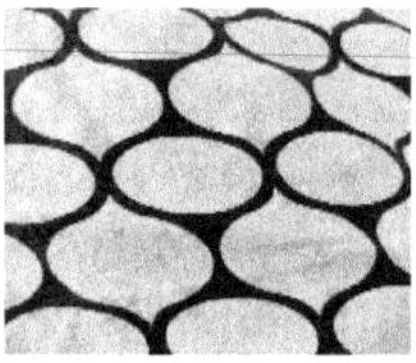

The idea of a star-shaped badge as a shield still exists worldwide, making police officers wearing it know that they are doing God's work, which means doing work in scientific reality to be humane officers, and to protect free self-thinking individual humane-being from believer ignorant followers of a leader in a inferiors gang.

PART THREE

Evolution

INTRODUCTION: SCIENCE AND NATURE

The truth about nature is that it requires a sense of dynamic equilibrium in the natural world. However, science accepts that all systems are in a constant state of flux.

An old saying: "If you give a man a fish, you feed him for a day. If you teach him how to fish, he will eat for a lifetime." Another outlook: "If you fix a problem for people, you will help people for a time; if you teach people to fix the problem, they will fix it for a lifetime." To that, I would add, "If you teach a person to free self-think in a scientific debate, he will create new knowledge to fix problems and will teach people to free self-think in a scientific debate to create new knowledge to fix all problems for a safer democratic joyful world. By challenging yourself, you will always feel good about your life and all lives, which is the path to the joy of living!"

Isaac Newton was a self-thinker in a scientific method of thinking. Newton saw an apple fall from the tree. He knew it was a reaction to a force that made the apple fall; it creates understanding of reality physics and science, and that's how he discovered the force of gravity. Gravity became a new word to name this further action. Newton did not know how gravity works. Albert Einstein discovered how it works in a scientific method of debating, research, in physics, and mathematics. **Billions of scholars with knowledge have seen an apple fall. However, they did not search for understanding in God's reality, because they are followers of illusion, and memorizing knowledge to become a Sheep in a herd of grazers.**

Newton also knew that the actions made by the universe's energy caused the reactions that created the universe and made life possible. He called that action God. God is a word used to name the unknown., Newton did not know how God works. Newton wrote about his discovery but had to hide it because he was afraid of being persecuted by religious leaders and their ignorant mobs of followers. It took 200 years after Newton's death for someone to find his book. Regrettably, today and in the past, some of our most extraordinary creative humane-beings were tortured and killed.

I do not blame anybody who wants to live in a better world. Sadly, people are still at each other's throats over illusions of truth from their beliefs and traditions. Some may not be willing to acknowledge scientific universal-truth, historical facts, verifiable conclusions, physics, and mathematics. However, these all contain universal truths that are not subjective to each person. It would be funny to see Einstein argue or compete or be violent with Newton about science because the universal truths they both discovered are not subjective to each person; they are verifiable conclusions. People fight over beliefs, religion, and tradition, and that creates the philosophy of the atheist! They are all not universal truths; therefore, they are subjective to each person because they are illusions of knowledge, not God's reality of scientific thinking. **It is time to stop believing in creating competition that creates fighting because it does not work in life reality or democracy**. Use your mind's resourcefulness by a scientific way of debating for a more accurate understanding of the spectrum reality of life to solve all problems.

CHAPTER 28

FREE SELF-THINKERS ARE PROBLEM-SOLVERS

I have a friend who likes to talk about scientific theory. It is fun to discuss subjects that we currently do not understand completely. One day we discussed deep space travel and how it could be possible to travel the great distances of the universe.

By using our imaginations and engaging in critical free self-thinking, in reality, we came up with some interesting theories. But, of course, theories are just self-understanding ideas; they are neither verifiable nor universal truths, unless proven as a universal truth.

When you electrically stimulate particles around an atom's nucleus, they jump from one orbit to another, faster than light speed. Nobody knows how this happens. However, it is a magic reality; so, I theorized that alien spacecrafts could change directions instantly and travel faster than the speed of light. This reminds me of those atomic particles that jump faster than the speed of light, from one orbit to another orbit.

I theorized that we should somehow imitate the particles traveling around the nucleus to jump from one orbit to another through space, simply. *But how?* Since orbits in a nucleus can be altered by jumping atomic particles.

The universe contains the magnetic fields of stars, planets, and asteroids traveling in orbit. These are like the concentric circles that occur when throwing a stone into a lake, creating rings. In our solar system, the forces of gravity create these permanent orbital rings in space. For example, the Earth gets caught in the Sun's magnetic spiral, which we call our orbit, while our Moon orbits in the Earth's magnetic circle. I share these thoughts because it demonstrates that you can develop theories that might eventually uncover new knowledge when you are free to self-think and use your imagination.

Science unifies us. The thirst for knowledge drives scientists to explore and discover new, universal truths. In the same way, it would be beneficial if more people aimed to solve societal dysfunction through scientific methods by free self-thinking. We would discover more answers and uncover more perspectives on subjects that will solve global problems. When everyone gets involved, we are more likely to succeed. As the saying goes, two heads are better than one. It would be great if people utilized their understanding of the search for universal-truth, to discover more ways to do things that will unify the world! When theories work, they are called "verifiable conclusions," a universal truth that leaves no room for illusory justification. That is problem-solving. Imagine no preachers worldwide, no followers, leaders, gangs, organized beliefs, hierarchy, criminals, lawyers, or judges, or jails, and instead of violence, we will have the freedom to free self-think in a democratic debate, searching to create a utopia.

My friend Dan always talks to me about space aliens. He says that they are all around us and have been around for thousands of years. He says that leaders of countries already know this and are worried about it. If this is true, then the aliens must be brilliant at traveling such long distances to this planet. The aliens must have scientists in wisdom that are smarter than ours. If so, this super-advanced outer space intelligence poses a threat to hierarchies all over the world. Hierarchies worry that aliens, if they exist, might usurp their powers by winning a war and they will lose their privileges. Government leaders tell their followers that space aliens do not exist, so don't believe your eyes if you see a UFO in the sky. Most preachers state that in the entire universe, only planet Earth hosts intelligent life. However, I am not a believer either way. I do not know if aliens exist or not. The hierarchies think they will have to fight the aliens, but only scared, ignorant, violent, greedy atheists leaders fight to protect their privileges whiz their blind-faith followers. Why not work with these intelligent visitors from the galaxies; if they exist, we can learn from them. Why would anyone have a problem with intelligent scientists in search of wisdom? I don't have a problem with scientists; I have issues with preachers of illusions for more privileges from greed to protect their basket of money.

If an alien astronaut space traveler knocked on my front door, I would welcome "him" and ask, "What do you prefer: coffee or tea?" I don't worry about intelligent aliens because I know that if we can communicate and carry on an intelligent conversation, we will learn from each other.

FREE SELF-THINKERS NEED ARTIFICIAL INTELLIGENCE

People use computers to speed up their work. In the late 20th century, Garry Kasparov was the best chess player globally. Still, at one time, the 34-year-old Russian was defeated by 9-year-old Deep Blue, a computer built by IBM with artificial intelligence. Many people did not think a computer could be better at judging chess moves or judgments in life. Still, computers are more accurate than people and have no illusions of knowledge, prejudice, hatred, ego, and fear of dying.

Since computers lack ego, they would be better than people to assess a case's facts and impartially judge guilt or innocence. That is why computers are used in mathematics, research and storing data. During trials, computers could replace judges, lawyers and jurors who may be biased, prejudiced, or create work where there is no work; and who may be influenced by their beliefs, not the facts. For example, some criminal justice-related employees accept bribes, some fear criminals, some are hanging judges, some enjoy playing God over others, some are religious believers, and some are criminals themselves, trying to get away with corruption throughout history.

On the other hand, Artificial Intelligence (AI) has no fear, no bias, no hidden agenda, and no sin. Also, AI doesn't compete for power and can be provided at a much lower cost to the taxpayer! AI does not have to be connected to the Internet, where it could be hacked by hierarchy. Using AI, people will get a fair judgment, not like the O.J. Simpson trial, a bad joke, and a travesty of justice. AI should be used because I do not know of anyone who likes judges and lawyers. People should have the option to be judged by a computer. This would also unclog the justice system in a democracy.

Jesus said, "He who is without sin, let him cast the first stone." Moses and Jesus knew that people were not perfect, so they did not want leaders, hierarchy, or judges. Today, Jesus and Moses would prefer computers to make judgments because they knew there are no judges, lawyers, or jurors without a sin. That is why I agree with Jesus and Moses that imperfect people should not be judging other imperfect people. For example, the Constitution was created to provide equal rights for all of us, under humane law. But women have more rights in this country than men because it is a billion-dollar business for divorce lawyers to represent men and women. And some lawyers create work when there shouldn't be. Some men and women get married to profit from a divorce. The result is, 50% of all marriages end in divorce or separation.

Meanwhile, most men and women in that situation conclude that marital sex was not worth it and was not a wise idea! Women should get 50% and men should get 50% of what they accumulated together by a computer. There would be no divorce for-profit, and couples would be motivated to stay married.

Today, there are more rich people in America than ever before. Ten percent of America's population owns ninety percent of the wealth in this country. With creative tax-avoidance loopholes and lobbyists pushing laws to secure more money and stocks for the wealthy, rich people created a system so their families will forever remain rich. That is a Republic not a democracy, it replaces royalty with rich people; they are "too big to fail." *Why did we fight the English royalty, then?* When you mention the Rockefellers, everyone will say that family represents American economic royalty. When you mention the red carpet in Hollywood, it represents media royalty, it is created by freedom of religious gang. However, you are what you do, everyone is unique. **Long ago rich people gave their land to the public, lick Griffith Park, and many other land, because America was a democracy, and in a democracy everybody should challenge themselves to be more realistic to today than yesterday, for a better life for all, that's what is by the people, for the people, from the people, not for creating parasite rich hierarchy,**

All great civilizations vanished because they expanded the rich, creating a world of gangs, royalty, parasite hierarchy, peasants, servants, slaves, and homeless people. That is why Babylonian hierarchy supports overpopulation to serve and sustain them in their luxury, reinforcing their belief that money, wealth, and power are the goals and God of civilization. However, overpopulation increases pollution, diminishes natural resources, contributes to global warming, and ultimately destroys our planet. The only solution is for the government to represent the people, to create temples of wisdom, and educate all the people on this planet, to free self-thinking in a debate to create more understanding the reality of life, to be able to comprehend beyond a doubt that there is a Creator of life Energy, to do what is right, and what is right is to be moral humane-being **with a hire consciences.'**

Religious leaders tell their flock that if someone is lucky enough to have mineral riches on their property or win a massive lottery jackpot, it means God loves them more because they are good obedient followers. Religious leaders tell these stories to entice followers to obey the illusion of religion and traditional law. The Universe's Energy created everything for life and God does not love some people more than others. Only a certain percentage of people have oil under their land or win the lottery. That is just a % (a percent) of people who win the lottery, not God's blessings upon you! It is not wise to have hierarchy judges with power over others. However, if the authorities treat people like sheep, they end up with a country of **sheep in the herd of grazers**. Many of these individuals will rebel and could be killed or incarcerated in costly jails because they are not sheep and are meant to free self-think in search of universal-truth from the day they are born.

My suggestion to ensure true justice is to have our computer geniuses devise a program that can judge truth, innocence, right and wrong, impartially and rationally, without prejudice, or ego, or belief from the illusions of knowledge. Use it to resolve all criminal cases and legal entanglements. After all, if a computer can detect a lie, or with an MRI, indicate whether he meant it or not and win a world's best chess player in a game, then a high-tech intelligent computer can determine the truth or innocence better than a person.

No laws should exist to protect the individual from himself or herself. If someone wants to harm themselves, that is their right to do so. But it is not their right to hurt anyone else. This is as fundamental as freedom of speech. Preachers say if you commit suicide, you won't get into heaven. Well, do we know if heaven even exists? It is your life, do what you want with it because no one can own you. No leaders or gangs have more rights than the individual, and we all are individuals first. The truth is that many scientific achievements were accomplished by humane-beings who risked their lives, such as Madame Curie, who died after discovering radiation, and Louis Pasteur, who saved millions of lives by developing the antibacterial process called pasteurization and the world's first rabies vaccine. They would never experiment with someone else. Such scientists used knowledge and free self-thinking to help life, and that's what humane-beings do.

Only accessible self-thinking humane-beings can create humane law! However, many rules do not make free self-thinkers because they are written by hierarchy, a law designed to remain in power. When innocent people are treated as criminals, it is due to too many laws: opinionated laws, religious laws, and unfair laws written to have power. All countries should be a country of humane laws, not a nation of laws made by kings, autocrats, religious leaders, or corrupt politicians. Capitalism is a balloon, and inflation inflates the balloon, eventually, the balloon will blow up.

It is like swimming: when you first learn to swim, you wear a life jacket and when you know how to swim, you will not wear a life jacket if you don't want to! However, the leader says wearing life jackets saves lives. Staying home saves lives too. **Being afraid to die or being afraid of reality stops people from living and stops people searching for universal-truth!**

Here is another ridiculous law: you must have a life jacket on a paddleboard or wear a life-Jacket while paddling, that will not work. However, if the board is attached to your ankle by a strap, you would not need a life vest because getting back on the paddleboard keeps you out of chilly water and it is a much safer flotation device than a life jacket. You have to make the choice that you should decide to stay safe, by freedom of the mind, not your leaders. That is why I invented a floating paddle with buoyancy equal to a lifejacket, so if you fall in the water, the paddle will keep you afloat. Leaders are fearful of dying because they have vast knowledge, but no understanding of the creator in reality of science, and therefore they write many laws. There should be only humane laws to have power over free self-thinkers; people who evolve into free self-thinkers need freedom. People are born and die every day. **Being afraid of hierarchical laws or religious laws, you become an obedient follower of orders that stop people from free self-thinking to become creative. Anyone that is meant to be born is also meant to die, that is why we have to become humane-beings, so we can understand the creator in reality of science.**

CHAPTER 29

WHAT IS LOVE?

Love of a woman, a man, a country, a race, a god player, a religion; or Jesus, or Muhammad, or Buddha, or King David, or anyone who seems so right. However, sometimes the feeling of love is just infatuation; it is not true love. It is an imaginary animalistic illusion if it does not come from free-self thinking accurate understanding. Sometimes infatuated love will make people act irresponsibly insane. For example, an infatuated person might love an immoral or murderer and still ignore all the evil deeds committed and inhumane. That is blind love, which causes people to lose their sense of reality and hinders them from understanding the way they indeed are, and that kind of love comes from animals in nature to create a better species. However, people by self-thinking in reality debate for the search of a more accurate understanding create healthy love. It is the love of yourself and all life and becomes a creative consciousness humane-being. Sex is great for a being, but healthy love is better, for humane-beings. People in illusory love get emotional and can't control their feelings. That's why if a person kills their spouse for cheating with another lover, some courts will accept an insanity plea. The same is true with terrorists who are willing to die for their **leader god player, who groom and dress well to look like a god.** My friend, Maurice Braslave, was a tailor he made you look like a god, his clothes fit perfect, they cost 50 times the price of regular clothes. He created clothes for Russian President Nikita Sergeyevich Khrushchev, Canady, Jerry Brown, and many celebrities. People are captivated by leaders' attractive looks and from believing illusory religion and are infatuated by or in love with their godly leaders' looks.

Musicians write songs and sermons in all religions about their love of all their religious leaders. King David's empire, became famous because of his emotional poems about friendship and God, creating a religion and tradition to unite a gang of obedient followers. This gives people a false sense of belonging, causing them to fall in love with something that makes them blind, because these songs and poems are preaching, and they don't create a self-understanding of reality. It is not healthy to love someone who preaches, that they love you but don't act like it by disrespecting or physically abusing you. On the contrary, healthy love is raising the other person. It is respecting and appreciating them. When two people have a healthy love, they must treat each other fairly. When one party only takes and the other only gives, it becomes unhealthy love. The greatest love derives from the wisdom that comes from an **accurate understanding of life's realities** as a humane-being!

Each person must be humane to accomplish a higher goal, such as raising children and caring for the elderly. It is like a business and each person must communicate and debate regularly, in reality, to do what is right. If a person does only what they want, without regard to other people's needs, that is not healthy love! Love without affection is a word without meaning. Dictators, royalty, and religious leaders keep taking and taking from their people, yet those same people still love their dictator, country, and religion. This is not healthy love, and it does not come from understanding.

Healthy love creates the most joyful, long-life force. Hate creates the most destructive sort of life force on this planet. Hate is created by belief, illusionary knowledge, tradition, philosophy, and propaganda, that just justify. If you want honest and kind friends, you must start with yourself. Always tell the truth, and speak up when you know someone is lying,

From an accurate understanding of love, love is in all living things, it creates a more beautiful species, and a beautiful world. For the love of all lives, and the joy of living, it is a good reason to be a humane-being.

The reason people sometimes have problems in their marriage is that they compete, and competition creates the ego. When couples or kids fight, it is because they want attention and affection from universal love. If they don't get it, they become unhappy and will fight. However, parents must communicate with everybody and teach them to think for themselves. You should give everybody the right to disagree.

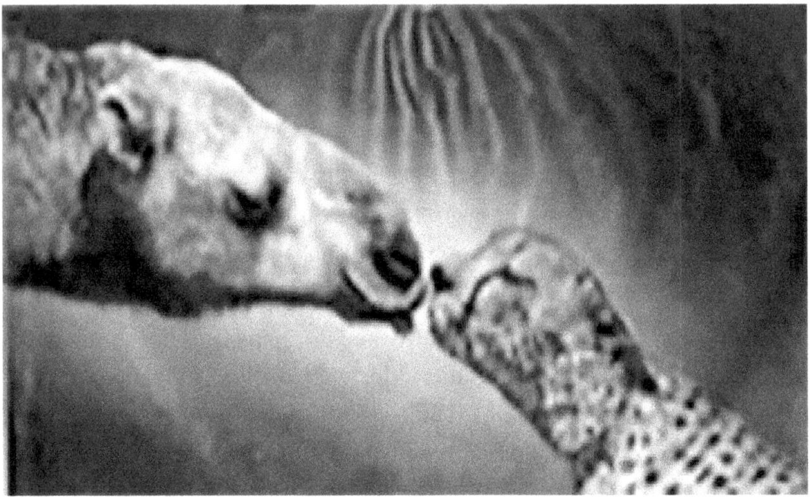

THREE STEPS TO REACH UNIVERSAL LOVE

Step 1 When you are born, you need the experience of being cared for, nurtured, and loved to get to Step 2.

Step 2 When you are a teenager, you need the experience of loving and being loved by a stranger to get to Step 3.

Step 3 Only then, can you represent any life. When you love your life and others because you have the joy to love all life.

If you are not loved or you do not love others, you will want to become a leader, by creating money, by hook or by crook; and by becoming a member of Babylonian atheist hierarchy, your followers or your family will have to love you.

CHAPTER 30

IS LIFE A GLASS-HALF-FULL OR HALF-EMPTY?

When we are born into this world, the glass is half-full. The half-empty part of the glass will be filled with future life-inspiring challenges. Maintaining this balance in life is believing their glass is half-empty; that person is blindly negative about life. If someone believes their glass is half-full, that person is blindly optimistic. Neither viewpoint is helpful or realistic because both are illusions of knowledge.

Babylonian religions propagate illusions to attract followers who think their glasses are full, so they can blindly believe that a. Santa Claus or whatever you believe in, in Heaven, will take care of everything. These preachers benefit the religious hierarchy, not their believers. Atheists who think their glasses are empty believe the illusion that there is neither a man, God, nor God of cosmic energy. Our understanding of life must be in scientific reality, so we can do what's right without delusions. Otherwise, people will become uncertain about how to proceed in life. **This is the reason people are Atheists, or criminals, or leaders.** However, people that free self-think and challenge themself to live humanely put themselves in the right place mentally. If you understand this, you realize that going against the humane way hurts others and hurts your mind. You have to take ownership of your actions by the scientific method of free self-thinking, without influence from leaders preaching illusions, and opinions.

One day at a coffee shop in my Los Angeles neighborhood of Silver Lake, a group of young gang members came and sat down at a table next to me. They asked what I was doing. I told them, "Writing a book." They asked, "What's it about?" I told them, "It's about how to free self-think to be creative," and then they asked me how much the book costs. I told them the price and they said they had no money. Then I asked them, "What is the most important thing in their life?" One of their responses was, "To observe policemen. I can spot cops a mile away. It doesn't matter if they're in an unmarked car. I will recognize them." I asked him why that was so important. He answered, “I have lots of friends in jail, and being in jail is no fun. So, I try to avoid cops at all costs.” Then I asked, “Okay, so what is the second most important thing in your life?” His answer was, “I make sure to go to church every Sunday.” I asked him why. He said, “Because I don’t want to go to Hell. I want to be certain I’m forgiven for my sins.” **I ask,** “Is not going to Hell more important than doing what is right in God’s reality of science here on earth?” His answer was, “I believe I have to attend church every Sunday because I don’t want to take any chances of burning in Hell.” Sadly, because of his brain's obstacles from believing illusions, he cannot understand reality to understand right from wrong.

Everybody plays the game of life; most people don’t know the game rules. Because it is legal to create propaganda untruths by selfish egotistic Babylonian hierarchy laws, and religious laws, created to control women and men, it never works and will never work for people, because it creates uncertainty, misunderstanding, homelessness, slavery, wars, and hate, and God is money. However, the scientific method of free self-thinking is to learn from others by realigning the brain into a more accurate understanding of the spectrum of magic reality to search for God's universal truth. To become confident, creative, humane-beings, that will make a better world for all lives**. It is time to make propaganda, religions, and all scams, not legal, to create a democratic world for all people!**

CHAPTER 31

WHY THERE BULLIES?

During my three years as a paraplegic struggling to walk again, I realized firsthand the abuses that occur when people with power prey on the weak and innocent to feel good about themselves. Mean-spirited kids from the neighborhood who did not have a humane upbringing by competition, were inferior bullied. They wanted to feel important from their power, so they thought it was funny to hurt an innocent child who couldn't defend himself. Because of my experiences, I developed a conscience that naturally led me to become a humane-being willing to fight injustice. And I still do. It is common knowledge that when people work together in a gang, it creates power. **The cause of bullying comes from being inferior and from the need to feel important.**

In ancient Rome, Christians living in the Roman empire did not consider the Romans to be humane-beings. The Romans did not tolerate criticism from the Christian minority, who made the Romans feel inferior. So, the Romans, from a competing ego, fed Christians to the lions for 700 years. This also served to keep other people fearfully obedient. Truthfully, the Romans were bullies and inferior. When Jews were a minority, living with Christians, they didn't consider the Christians to be humane-beings, so the Christians, for an ego, tortured and killed the Jews during the inquisitions. Truthfully, the Christians were a gang of bullies and inferior.

Today the Christians and the Israelites don't consider the Muslims to be humane-beings, so Muslims, from an ego, become a gang of bullies because they feel inferior. Even in school, successful kids don't think other kids are good enough to socialize with them. So, the kids who feel inferior, from an ego become a gang of bullies. Since the beginning of time, free self-thinkers have been geniuses. That made the world a better place, from an ego, and they are tortured and killed, because they are a minority. Today we know and understand that we need to be a majority of free self-thinkers to save the world from the United Nations (united hierarchy), because we can destroy the world with guns, bombs, atomic weapons, poison gas, and epidemics. We must stop, learn, and teach how to be free self-thinkers in wisdom to realize that bullying is not the answer to feeling better by competing with an ego from powers. **It is time to humble yourself, to learn from each other, the magic of God's reality of science, to find universal-truth in a temple of wisdom and become peaceful creative humane-beings. Therefore, competing is not the answer for people's futures!**

All free self-thinkers continually exercise their minds in scientific debates, which helps them live happier, longer lives through an accurate understanding of reality. For example, Abraham and Moses lived to be over 120 years old. In the case of Steven Hawking, the doctors told him that with his crippling illness, he had only two years to live. Nonetheless, Steven Hawking lived a long life by free self-thinking in the search for universal truth! They all broke the law of doctors and embraced the law of God's reality of science and became humane-beings that create the love of life, to make you live a long life with a smile, and with hate from religion you will create a miserable short life!

How do you know if someone is a psychopath? If someone says all people are rats, that is a psychopath. If someone says if all people disappear from this planet, it is okay because nature will be better off, that is a psychopath. However, among people, humane-beings will protect all lives and the environment, from understanding, not from knowledge!

One sunny day on my walk, I met a man on a bicycle, the man said, "Why don't you come to my church, to be saved?" I said, "Why don't you come to the Public coffee shop of wisdom, to understand for yourself, God's reality of science, because God gave everyone his own brain, so they can free self-think to understand reality for themselves." The man said, "You are going to hell." I spoke. "You go to hell!" The man said, "Why are you insulting me?" I said, "This is the problem, when people are ignorant obedient believers in blind-faith addiction from lies, they don't realize that they insult, kill, and get killed, creating hate and war over illusionary belief! That is what the word stupid means. And as long, as people are sheep from being ignorant from believing illusion, they will not understand right from wrong, or God's reality of science for a meaningful, peaceful, joyful life!"

The Inquisition: what a show.

The Inquisition: what an inhumane blow.

The Inquisition: hate and war is here to stay.

Because religion blind-faith addiction is hard to go away.

All gangs give the wrong perception that creates an addictive illusion of knowledge. These gangs include religious organizations, atheists, the mafia, street gangs, royalty, and more.

Gang is not a personal thing; it is a follower thing! They all want to keep their power, throne, money, subjects, slaves, and followers. Eventually, they will die without understanding God's scientific reality. This time, they are too intelligent to fail and continue destroying lives and the environment. In history, they all fail because God will not have it. This time, people must free self-think to become the majority of the world's population and by debating to arrive at a universal truth, and that will unite all the people and solve all problems, to create a joyful utopian life, at that point.

Babylonian hate and war-makers' religious leaders will run away from their dangerous followers believers, when the followers discover, religion was all a lie, for mor money!

DECLARE YOUR BRAIN AND MIND INDEPENDENCE

When someone asks me, "Who are you?" my answer is, "I am a free self-thinker." Their next question is, "What do you do?" My answer is, "I search for universal truth for the time, by asking questions to understand the meaning of what is right in God's reality of science, so I can make up my mind to do what is right!"

When someone is directly confronted with the ignorance of their most cherished beliefs, they will likely become angry and defensive by clinging more tightly to their beliefs. Thus, you are facing them with a cognitive dissonance that will result in extreme discomfort. Because they cannot comprehend life without believing it. In the manner of Socrates, it is more productive to ask them questions about their beliefs and let their answers reveal the truth. They will ultimately come to a position that will conflict with their beliefs, by creating a more accurate understanding. Then they will either choose to stay ignorant or declare their mind's independence!

Therefore, I declare myself a free self-thinker; this is my brain and mind independence declaration. When I don't know something, I say, "I don't know," because the truth is that I just don't know. That is why I came to America to be free of the ignorant illusion of knowledge from all inferior majority gangs, that protect Babylonian hierarchy privileges. I worked hard, bought a home, established myself as a free self-thinker, and paid my taxes. I don't want to deal with any aggressive religious or ignorant gangs, bullies that cannot handle the universal-truth and intimidate me or anyone else to obey and follow them and tell me or anyone else how to live by violence.

Free self-thinkers that debate in the search for universal truth do not mind being questioned, and don't create propaganda because they are searching for truth!

CHAPTER 32

MY VERIFIABLE CONCLUSION REGARDING MOSES

After researching and debating the original ancient books about Moses from the island of Djerba , I learned that he was the first person to say that God was energy. Before he liberated the slaves in Egypt, he went up to the volcano, Mount Sinai, which was rumbling at the time, and he promised the life energy of the universe that he was going to end slavery in Egypt and the world because he was a humane-being. Moses was the Pharaoh's oldest son; he understood that God was Energy because he was surrounded **by scientists who debated in God's scientific reality for the search of universal understanding, created many scientific wisdoms in astronomy, built the best chariot of wars, built the pyramids and more!**

Moses was the first son of the three sons of Pharaoh Akhenaton. In those days, some Egyptians were brilliant astronomers and scientists. They understood that Energy is in everything and understood that Energy can neither be created nor destroyed. Unfortunately, and it has happened throughout history, and still today, the free self-thinker scientists under Akhenaton did not understand the dangerous cruel leaders and their religious ignorant followers in blind-faith, who are willing to kill and get killed to protect their lies, believing that they are important.

That is why the scientists were surprised and killed by religious leaders and their majority of ignorant followers because leaders wanted to rule over Egypt to protect their privileges. They wanted the people to keep believing the old religion of many gods, and not to believe the new scientific understanding that Akhenaton, the Pharoah, wanted to establish from polytheism to monotheism.

Akhenaton understood from scientific research that there is only one God, one Energy, that correlated and encompassed everything in the universe; he called that god-energy **Aten.** Secretly rebelling against Akhenaton, the religious leaders had Akhenaton poisoned to protect their power, money, comfort, and privileges. Because Akhenaton wanted to change the illusory Babylonian religion to a scientific reality way of thinking**. The goal of all religions was and is to keep followers in blind-faith to stay idle and by violence prevent the scientists from searching for universal-truth in God's reality of science.**

Since he was a boy, Moses knew that he would inherit the Pharoah's throne someday as the oldest son. As a young man, Moses had to escape to the desert because he knew the priesthood would have assassinated him as they had assassinated his father. Because Moses understood his father's scientific cosmic energy and had a scientific understanding of wisdom to unite homo sapiens to speed up evolution. He worked for a shepherd for a few years until religious leaders' violence calmed down, then Moses returned to confront his younger brother, the Pharoah, and demanded his right to the throne because he was the oldest son. However, his brother, the new Pharaoh, made a deal with Moses. Moses would inherit the slaves because they were the most ignorant, the lowest caste of society, and his younger brother would stay in power as the Pharoah.

The former slaves were a mix of people who had been defeated at war. Some were Hebrews and others were from North Africa or West Africa. They were the people given to Moses to establish his free self-thinkers to understand God's life Energy. Moses understood that even slaves from the lowest class of society could be brilliant humane-beings if they become individual free self-thinkers in a scientific way of debating for accurate understanding, because nobody is born important.

Before Moses, Hebrews believed that God was some sort of a being or an entity. After Moses, the Hebrews understood that God was Energy because universal truth changes with time. As per Akhenaton's scientific conclusion of God, Moses created a humane law - 10 Commandments with a piece of charcoal from burning bush energy. It was the first Constitution between people and God.

His goal was to unite all the people of the world and speed up their evolution into free self-thinkers humane-beings and become enlightened to solve all problems. However, today, for a more accurate understanding, you should obey only 7 Commandments and 6 Days for people and animals of burden to have the seventh day for rest, and 6 Years working the land, and the seventh Year is for rest after working the land. These laws are still used today.

The truth is, no person can carry two marbles of stones, one on each arm, and go down the mountain, because leaders create more laws, and change books. **People don't need too many laws to become creative humane-beings:**

1. Moses' first God commandment. **'I am your God, don't create other gods before me, not to worship leaders and preachers (god Players).**
2. Don't compete for an ego, instead challenge yourself to be better today than yesterday.
3. Free self-think in a debate and learn from each other to search for the universal truth to unite all life, for a safer and more creative world.
4. Don't lie.
5. Don't cheat.
6. Don't hurt or kill others, unless defending yourself.
7. Debate with your father and mother to create a more accurate understanding of the magic of God's reality to be a free self-thinker humane-being.

If you don't, your father and mother will create a family Hierarchy that will create religious laws that create traditional laws to believe that you are important from your ancestors' illusionary belief, in blind-faith addiction!

Moses was willing to suffer in the desert because he was a humane-being searching for wisdom. He did not believe in God, he understood God. He knew right from wrong and challenged life to make a better world for all lives. If Moses had been a greedy leader, he could have led a comfortable life as a Pharaoh. Instead, he spent forty years in the desert, teaching slaves to understand reality better, because, it is not what God can do for you, it is, what you can do for God!

Moses was brilliant; he knew the criminals who poisoned his father were the religious followers and their leaders who were still harming others who did not join their religion. Moses even knew these criminals intended to kill him, for power over the people to create privileges.

Moses realized that ignorant religious followers were scaring people with violence to create more followers and enhance their religious gangs' power. Therefore, Moses asked free self-thinking people to separate from religious people in tradition, to keep the peace. Then he went up the mountain with his army and waited for them to separate. After they separated, Moses and his army came down the mountain and massacred every religious person and their criminal leader. In the books from Djerba, it says Moses killed 1500 of these religious people! Some people don't like the part that says Moses killed religious people.

Today's religious leaders explain that they were swallowed whole into the earth's bowels when the ground opened beneath them because they worshiped lies and illusion idols**; however, if you can accept that then you must accept that all tradition from religious people will be swallowed whole into the earth's bowels when the ground will open beneath them because they worship tradition from religion's illusionary lies that keep people idle, as their children repeat the same.** Moses wanted a world of free self-thinkers, creative humane-beings, scientifically debating in search of God's universal truth, in a democracy.

The tale about the infant Moses floating down the Nile River in a basket is a make-believe fairytale made up by religious leaders to attract believers with their "basket-case" stories of illusion. All religions share a "basket-case" fairytale that distorts reality, but makes their followers feel important and know that a **cruel god player** loves their gang of believers more. Moses said that when you pray, pray to the universe's energy, not **(god-players) or angels.**

When Moses was dying, he did not want anyone to worship him or his grave. So, he went up the mountain by himself to die alone. It wasn't easy for someone over 100 years old, but he wanted to show that no man should worship any man or woman (god-players). So, as a result, even to this day, no one knows where he is buried. Because if they did, ignorant religious people from all over the world would go there, to worship his grave and create propaganda, to make believers believe that they are important. You cannot change hieroglyphic writing, caved in stone; however, leaders can change books, destroy books, and create "basket-case" storybooks to create obedient believers in blind-faith addiction to feel important

Moses never fought any wars because he wanted people to stop killing each other, which became part of the commandments law, "Thou shalt not hurt, kill." In wisdom." Violence is for animals and scientific debate conversation is for higher consciousness humane-beings. You should check it out for yourself.

Today, archaeologists, governments, religious leaders, and scholars praise the Pharaohs (god-players) for their power over others and wealth. However, for all their elaborate burial chambers filled with mummies, gold, silver, gems, and other precious items, none of these so-called Pharoah gods achieved actual immortality, except perhaps in history books. Still, in reality, they are like dinosaurs, dead as doorknobs. But, on the other hand, Moses and his father, Akhenaton, achieved true immortality in the hearts and minds of humane-beings worldwide, which inspires new generations of free self-thinkers and creative humane-beings around the world, and that is the reason, we are, mor advance in science today. Thankyou Moses!

Years later, those ignorant slaves of Moses evolved into free self-thinkers who've achieved incredible milestones, including 25% of the Nobel Prizes awarded for science alone. These descendants of wisdom are only 10 million out of over the 10 billion people on the planet. That is why Hebrews' free self-thinking is more effective in making a better world than following a religious tradition from hierarchy. Akhenaton's and Moses' wisdom movement which achieved the verifiable conclusion that God is the Life Energy of the Universe still exists today in scientific reality. The rest is in history! The result was the end of Pharaoh's god-players. The end of the Greek god-players. The end of the Roman god-players. The beginning of humane democratic law that sped up our evolution by **free self-thinking democratic scientists. Thank you, Moses**.

My verifiable conclusion is that Moses gave his people the ability to free self-think in scientific understanding. However, his brother, the Pharaoh, led his Egyptian people to believe in religion, tradition the illusion of knowledge. He was a god-player. As a need for obedient followers, all the questions are in history and the present! **This leads to the critical questions: *Was Moses' 40 years challenging? Are life and people***

suffering in the desert without comfort worth it? **For humane-beings in wisdom, you bet it was!** Evolution shows that neighboring countries use words, and ideas from other countries as their own. For example, **democracy** in Latin is, **De Moses,** meaning in English, **from Moses; democracy** in Greek, meaning in English, **ruled by the people! Democracy unites all the people, and leaders separate people!**

If you let religious leaders explain the meaning, by preaching about Moses you will be addicted to believing in religious illusion, and tradition in blind-faith; and blind-faith is a distractive gambling addiction. It is considered in the medical field as a sickness disease that is hard to cure by believing that you are a winner. However, if you have a gambling addiction you are not a winner, you are a loser! That is the reason Moses spent 40 years in the desert, to cure people of addiction in believing religious tradition, in blind-faith. However, curing people of slave mentality takes a second. All you must do is pay for their work!

MORE OF MOSES' INSPIRING WISDOM STORIES

When Moses escaped into the desert as a young man, he met an old shepherd, who hired Moses to lead a flock of sheep. The older man proposed that Moses's salary would receive all the lambs born with two colors. Because he knew that very few lambs are born bi-colored, he naturally assumed he would get most of the lambs. So, Moses painted sticks in two colors and placed them around the sheep's waterhole. He understood the power of the brain, that if the pregnant sheep watched those two-colored sticks while drinking, it would affect their lambs' color. The result was that most lambs were born bi-colored, and Moses now possessed a flock of sheep. From understanding, not from memorizing knowledge, he became rich. That is why women should not experience stress or watch scary movies while expecting. Moses certainly had a more accurate understanding of the magic of God's life reality of science.

Moses didn't want people judging others because that created "god-players." He knew people make mistakes and nobody is perfect. So, Moses built a box to hold the 10 Commandments, the contract (the constitution) between his people and the Creator, the Energy. The concept was God of Energy provided the earth's Goldilocks' atmosphere for all life to flourish. In return, Moses' people promised to evolve into humane-beings to serve as an example for the future to unify all people. Moses was brilliant from an accurate scientific way of thinking; his wisdom changed all people for the better by speeding up evolution and keeping scientific thinking alive. However, hierarchy creates all religions, and religion makes tradition to create obedient followers; obedient followers create a gang. That is why Hebrews created a hierarchy to defend themselves from other gangs, even if they did not approve of hierarchy or gangs. They did it to protect the Hebrews from gangs that kept attacking violently in natural law, created by cruel atheist leaders.

CHAPTER 33

MY UNDERSTANDING, FROM DEBATING THE STORY OF JERICO

***Jerico's walls came down, from the inside out.* The representative of the people of Jericho did not like their cruel inhumane hierarchy (god-players) and preferred the realistic God, the Creator's life Energy of the Hebrews. So, the sound of the Horns was a signal for the people of Jerico to revolt, and they united with the Hebrews because Moses wanted to unite all the people of the world in a democratic God's reality of science. It was all debated in an earlier Hebrews' trip to Jerico.**

Babylonian religious leaders explain in religious books, the walls came down by a miracle, so the Jews could come in and conquer Jerico by a violent war because there is no bucket of money in peace for Babylonians leaders. All religious leaders in all languages use the word blasphemy, to scare people about God. However, why should those religions worry about protecting God from what someone says? What kind of an all-knowing, all-powerful God would need our protection? It is irrational to think that any God would need our protection. However, a Babylonian king, pope, or leader (god players), will someday need protection from their dangerous illusionary obedient followers, when their followers will find out, it was all a lie, despite their continuous preaching, by using "the guillotine! And this is what is happening now! 2 countries are no man land, because Babylonian religious leaders are killed or escape fearing their cruel followers. But it is rational to think that all life needs representative humane-beings' protection from dangerous inferior believers bullies' and obedient followers. The Cohen Hebrew people are created to be the representatives, for all the people of this planet in a democracy because they are chosen to be the most humane.

CHAPTER 34

KING DAVID

The 12 tribes of the creative Hebrews have been continuously attacked by inferior bullies, the Philistines, who created a dispute over a piece of land. To settle the dispute, 2 warriors will fight, and the 1 that wins will get the land. So, the Philistines presented a powerful warrior, Goliath, and the Cohen Hebrews had wisdom. They presented a Shepherd that can kill a wolf from a long distance, and David won the fight. However, it was not a fair fight. David was an attractive young man with curly red hair and blue eyes, so people voted for him, and he became a king, hierarchy.

The hierarchy made a statue of King David because King David looks like a God, and he created a powerful empire and that was the symbol of hierarchy. King David conquered 10 times the size of the Hebrews' land and inherited the Babylonian hierarchy into his kingdom, because King David's followers, following him in blind-faith, were willing to kill and get killed because they were stupidly blind. **King David was from the Juda tribe, and that is why the Babylonians created the word Jew because the word correlates!** One sunny day a beautiful woman by the name of Bathsheba was taking a shower, naked on the roof. King David saw her, so, he wanted to have sex with her. She said that she was the wife of is general. King David had sex with her anyway, he had the general killed and promised Bathsheba that her son, Solomon, would be the next King.

King David was and is an example of Babylonian world hierarchy. Because there is no profit in peace for hierarchy, and hierarchy must create hate to unite, they are needed by ignorant obedient followers. **However, in reality, hierarchy and leaders are not needed in a Hebrew Democracy! That is the reason 10 tribes, out of 12 tribes, left the land of Israel to create a democratic world and reunited back, in the land of *Israel, as* free self-thinker Hebrew Zionists, for that is the reason Hebrew Cohens left the land of Israel, 2,500 years ago to the island of Djerba and come back to reunite with the Hebrew Zionist, fleeing the Babylonian magical show that created the religious Jews!**

CHAPTER 35

WISDOM OF KING SOLOMON

King Solomon was another free self-thinker in wisdom. He could have lived very comfortably as a king, son of King David; nevertheless, he understood Moses's wisdom and the importance of the Hebrews' humane law. **King Solomon worked every day as a judge without judging people.** The question is, how? He devised ways of making the accused judge himself by creating their own understanding of right from wrong; that is why the guilty party understood he was guilty and became humane. King Solomon understood the importance of respecting brilliant, creative working people. So, he wrote the Book of the Masons, and the Masons wrote the American Constitution. King Solomon left all his fortune for the Masons and left nothing for his children who wanted to be hierarchy. He was the first to create the missionary movement that helped produce more humane-beings, he wrote the Golden Rule **and taxed heavily the Babylonian people that competed to become hierarchy. They were from King David's empire,** which created religion and blind-faith addiction that is hard to cure to become god players, and they treated money as God. King Solomon succeeded in bringing back the free self-thinkers, who were the Hebrews' democratic scientific movement of Moses, that called themselves Zionists to unite all the people of the world in search of new scientific universal-truth.

In King Solomon's realm, a restaurant owner who worked hard for his money came before Solomon's court. The owner complained about a beggar, who stood outside his restaurant every day to smell the food while eating some bread he brought with him. The beggar never spent a penny inside the restaurant. Because of that, the restaurant owner took the beggar to see King Solomon's court to be tried. King Solomon inquired, "How many times did this man smell your food?" The restaurant owner replied, "A whole year's worth, over three hundred times." The King asked, "How much does a meal cost?" The man replied, "One dinar." Then King Solomon told the beggar to give the court guard one dinar. The beggar said, "But that's the only money I have in the world!" The court guard took the dinar anyway. Then King Solomon instructed the guard to drop the coin three hundred times on the floor and then pick it up. The guard did so. Then King Solomon said, "Now both of you are even. This beggar smelled your delicious food three hundred times, costing you nothing, and you heard the defendant's money hit the floor three hundred times, also costing you nothing. No one should create work where there is no work or get something for nothing to become a leader." The guard returned the beggar's dinar and that is what was humane justice.

Another time King Solomon displayed his powers of reasoning was when two women on the same day gave birth to one baby, each in the same barn. During the night, sadly, one of the newborns died. Before the other woman woke up in the morning, the grief-stricken mother of the dead baby decided to place her deceased infant next to the sleeping mother and take the living baby as if it were her own. In the morning, the living infant's mother realized that the dead baby next to her was not her child. The two women quarreled over whose baby it was. So together with the infant, they went to King Solomon's court for a final decision as to who was the actual parent of the living child. Solomon ordered his guard to take the baby from the women. When he did, King Solomon told the guard to slice the baby in half with his sword. In that way, each woman could have half a child. As the guard was about to cut the infant in two, one woman remained silent while the other woman screamed out, "Stop! Don't kill it! Give the baby to her." This proved to Solomon who the birth mother was because she was willing to give up her newborn child to live.

Solomon did not have to judge who was the rightful mother because the mothers judged themselves. The point is people have to judge themselves by knowing right from wrong from their accurate understanding of the spectrum of life reality to be humane, for wisdom! Here is another example of King Solomon's wisdom. The King of Persia asked King Solomon, "Why are you the smartest man in the world?" King Solomon replied, "I neither practice nor preach religion, **my way is democratic search for universal-truth, a scientific movement to create wisdom to protect this planet and all lives. Otherwise, what are we here for?"**

The King of Persia told Solomon that he would like to learn more about his movement, to which Solomon asked: "How would you like to learn, sitting down or standing on one foot?" The King of Persia said, "I don't have a lot of time, so I'll learn it, standing on one foot." King Solomon responded, "The condensed version of the movement is to understand the wisdom of humane-law: Do unto others as you would have them do unto you. This is the Golden Rule. It is the wisdom of humane law; it is not a religion."

King Solomon, the author of the Book of Masons, wrote that God the Creator of Life Energy of the cosmos is the greatest of all builders, making the stars, the galaxies, the Sun, the moon, and our planet Earth with just the right ingredients for all lives to flourish, including people who after all are the children of that universal power. The Book of Masons is a humane law. The condensed version of the book is **that people are builders and creators who are godlier than those who do nothing or seek power over others to create a parasite hierarchy. Because this planet belongs to all life, you will lose it if you don't use it or if you abuse it.** It was not meant to create a religion or tradition. It was meant to create a higher consciousness as a responsibility.

King Solomon sent missionaries into Africa and Asia to teach humane law. That is why some Black people in Africa today are Hebrews. To be involved in the scientific movement of free self-thinking and be an example of a creative humane-being. The Roman Babylonian Christians copy King Solomon and send missionaries all over the world to teach religion for more power for leaders. King Solomon was ahead this time but failed because he underestimated the powerful organized hierarchy and their majority of blind-faith believers!

As the Energy of the Universe would have it, thousands of years after King Solomon's death, the Night-Templer Crusaders found the Book of Masons and continued King Solomon's movement of wisdom. The Masons built more than 8,000 beautiful structures and churches in Europe and around the world. They built the Vatican and other churches, and the Louvre for the King of France. They started the first banking networks in the world to help free self-thinking creative people, not as a profitable business. The Book of Masons helped shape my understanding of fairness and humaneness. As a result, I have learned to ensure honest representatives of the people, and the people must become good at asking tough questions and not merely an obedient believer that leaders can scam. Unfortunately, people seek power over other people by competing to be more important.

What I learned about John the Baptist and Jesus is that they were also incredible free self-thinkers. They understood Moses and King Solomon's wisdom. Jesus said I am from the King Solomon movement. John and Jesus understood the importance of teaching people the wisdom of fairness to know right from wrong. They even gave their lives, suffering torturous murder, because they understood people's highest achievement is to become humane-being. They did not focus their lives on being rich or powerful.

What I learned about Martin Luther King Jr. is that he was a tremendous free self-thinker. He also risked his life to teach people how to be humane-beings and unite us as a democratic country to be an example for the rest of the world. Unfortunately, throughout history, many instances in which geniuses, who stood for the universal truth of humane law, were assassinated or tortured because they happened to interfere with existing religious, or tradition in blind-faith Babylonian hierarchy statutes.

Laws must be based only on the wisdom of "humane law." This is a simple system based on a scientific thinking method that creates a universal-truth humane law. Treat others the way you wish to be treated and free self-think like all folks must do, to create a more accurate understanding of God's scientific reality, to do what is right. The United States legal system was established to decide what is or what is not tolerable. In a melting-pot society that welcomes all groups from all backgrounds, we must be keenly aware of the legal system's social impacts. In a community such as this, tolerance is necessary. However, should we live life merely to tolerate other people and their beliefs, just so that our own beliefs will be tolerated in return? The unfortunate thing about democratic nations like the United States—and others in the world, don't correlate with religion, because gangs of one belief or another encourage their followers to overpopulate to eventually gain a majority to win elections and get their way.

Today, lives are in the hands of warlords, tribal leaders, business tycoons, dictators, kings, and religious demagogues. Even more disturbing, there are over one million slaves.

CHAPTER 36

ABRAHAM

Abraham was very wealthy, with plenty of land and livestock. He was a hard-working free self-thinker; **he understood that he could not have children, because he was not able to be clean at that time**. That is the reason he circumcised himself, and Abraham was not a Jew, because Juda did not exist until Jacob had 12 children and named one of his children Juda. **The followers of Abraham are Hebrews, because word correlate, and the 3 faders of the Hebrews, are Abraham, Isaac, and Jacob**

Abraham was 100 years old and would surely die soon. One fateful day, Abraham was ordered by religious leaders. God demanded sacrificing (kill) Isaac, his son, to prove Abraham's obedience to God (god-players). Isaac was the only living heir to his fortune. Abraham realized that the religious leaders wanted his wealth; he had witnessed preachers attracting followers by saying they spoke with God. He knew that it was false because God did not use words or demand sacrificing people. After all, since words can have several meanings, words are not accurate, often leading to misinterpretation and God, the Creator is precise. Otherwise, it does not work in such perfection. Many leaders have been cruel throughout time, but the truth is that cruelty does not come from God or reality. It comes from egotistic animalistic psychopathic atheist parasite leaders. This is an example of how he discovered the wisdom to diffuse the situation by free self-thinking.

Since Abraham knew no one could talk to God or vice versa, he realized the religious hierarchy simply wanted to eliminate his only heir so that when Abraham died, they could steal his animals, property, and wealth.

Wise old Abraham realized he had no choice because a free self-thinking minority cannot fight a gang of majority ignorant obedient followers. So, he told the priests' religious leaders that he would sacrifice Isaac himself. They went up the mountain, Moriah, together with Isaac, carrying the wood to be used in the sacrifice. Isaac was in his twenties and indeed he could have defended himself from his father if he had wanted to;

however, Isaac knew it was all a religious act. When they finally reached the summit, Abraham built a fire and bound his son. Isaac allowed himself to be placed on the altar, which shows that Isaac continued to trust his father's free self-thinking wisdom.

When they came off Mt. Moriah, he told the religious leaders what happened. Abraham said he was ready to sacrifice Isaac, but God spoke up and said to him that "because you were such an obedient follower, I will spare your son. Here is a sheep instead," which they promptly cooked and ate. I'm sure they were hungry because it took them three days just to reach Moriah's summit.

Abraham's story's wisdom is that Abraham appreciated his son's willingness to trust his father's free self-thinking wisdom. The religious leaders appreciated Abraham's apparent obedience. They used that as the basis of a story to keep other followers obedient, too; Christians, Jews, and Muslims still tell this story today by leaders to foster obedient believers in blind-faith. In the end, Isaac lived, Abraham died a natural death, and the religious leaders got a good story to tell about the way lazy ignorant followers should obey their leaders. Abraham succeeded in fooling the hierarchy because he was a free self-thinker Hebrew, not a religious man in a tradition. Since most people were believers, some free self-thinkers began following him because Abraham was a humane-being and a representative of the people; he did this to save them from being subjected to religious cruelties. By wisdom, he united people who helped to make the world a better place. He confronted the leaders of his time because he understood right from wrong. This is how wisdom creates a better life for all, and if you think that God demands to sacrifice people, then you are ignorant from believing illusionary knowledge and out of your mind. You are into memorizing what others have written to create obedient followers from believing religion that creates blind-faith, without free self-thinking in a scientific debate, for a more precise accurate understanding of the reality!

Today, religious leaders and some scholars say that aliens built the pyramids. The pyramids are amazing because brilliant people with a scientific method of free-self-thinking made them! The pyramids' food is still good to eat after 4,000 years because the Egyptian free self-thinker scientists understood energy better than today. We don't even understand the full spectrum of Energy in the Universe.

CHAPTER 37

HIERARCHY AND JOAN OF ARC

Despite tremendous odds against her, Joan's mission to save French Catholicism from the English Protestant army was victorious. She led the French army to a "David vs. Goliath" victory over the English. The news of her victory quickly spread across the British Isles and all of Europe. However, defeat at the hands of a teenage French girl gave the English hierarchy a black eye and a diminishment of their power and comfort to which they were accustomed. The English hierarchy retaliated by spreading the false story that Joan was an instrument of the Devil and must be killed. This is an example of hierarchy propaganda lies, to stay in power.

Even though Joan's victory saved them, the French royal court distanced itself from Joan. The king and his advisors became jealous and afraid of Joan's popularity among the masses. The French hierarchy did not want Joan to become more politically powerful than they were. After another battle with the English and their French allies, Joan was captured. The French king did nothing to negotiate for her freedom.

The Christian Catholic hierarchy was very angry at Joan, who represented a dangerous threat to their power. People were taught to "pray, pay, and obey." When Joan first revealed that she saw visions and heard the voice of God, the Church leaders realized she was a significant threat to their power and comfort. The Church had taught its followers that no one could directly talk to God except the Pope. Worshippers could only access God by going through the Church—and paying for it. Joan's statement about directly communicating with God threatened the Babylonian hierarchy's positions and income for doing nothing, and they decided to eliminate Joan. Imagine this for yourself.

After Joan was captured, the English and their French allies trumped up multiple charges for a trial against her. None of the charges or standards for an impartial trial were met. During her imprisonment, she endured numerous death threats from the hierarchy. Finally, after a year in captivity, the 19-year-old Joan gave in and signed a confession saying she had never received divine guidance. Joan was convicted of heresy and sentenced to die by burning at the stake. When individuals go against the Babylonian leaders, religious monarchy, and their ignorant obedient followers in blind-faith, it is hazardous and deadly to be a free self-thinker!

Joan was a representative of the people for the people. She upset the status quo. Her enduring success is that her story has encouraged other people to explore a direct relationship with God, the Energy of the Universe—without Babylonian religious leaders or monarchy. Joan did not survive her interactions with the hierarchy. Her enemies had condemned her to die long before her capture. She wanted to experience life because she knew how very precious and gratifying life is. But it didn't matter. The authorities publicly burned her alive at the stake and then later made her a saint.

The Statue of Liberty was given to the United States by the French. It represents Joan of Arc with the Torch of Freedom lighting the way for people to escape the tyranny of religious hierarchy by free self-thinking in a debate to create a more accurate understanding of the spectrum scientific reality of life, to be creative, and make a safer, democratic, loving world.

The French Revolution happened because Mary Antoinette, the queen of France said, "If you don't have bread, eat cake." So, when they were going to cut her head off, they asked her, "Why did you say what you said?" Mary Antoinette said, "This was my belief and my opinion at that time, I am sorry. I did not try to understand." They said, "Sorry, it is too late." So, please, try to accurately understand scientific reality, before it is too late. No excuses allowed!

◇◇◇◇◇◇◇◇◇◇◇◇◇

Here is a wisdom story from a free self-thinker: a driver got a flat tire in front of a mental institution. He stops the car and takes the tire to a mechanic to fix the flat. When he returned, he could not find the four bolts needed to attach the repaired tire. The driver was upset because he would have to go back to the mechanic. However, one of the residents from the mental institution who was watching said, "Why don't you take one bolt off each of the other tires and install those three bolts on the fixed tire to drive to buy replacement bolts?”

This is an example of how anybody, by free-self-thinking without memorizing knowledge, can be brilliant, even a mentally ill person, because God the universal energy is in you and all around you.

PART FOUR

Going Back Home

CHAPTER 38

WISDOM STORIES ABOUT THE ISLAND OF DJERBA

All ancestors' Inspiring Stories are just wisdom stories, and by debating you will find the original meaning and the reason it was written, not for believing in blind-faith. My grandfather also said, "Grandson, we escaped the Holy Land to the desert island of Djerba 2,500 years ago, fleeing the Babylonian magical show that created the Jews, and religious law to give power and privileges to leaders. For example, the Jews, believe, that the Tora was given to the Jews by God, (god-players) but my question is: Was the Tora, delivered by UPS or UBS? The truth is the Tora was written in Babylonian Aramaic, not by the Hebrews in the Hebrew language, and their magical God show was better than the Hebrews' scientific democratic universal-truth, God, the universal energy.

The Babylonians destroyed King Solomon's temple of wisdom to become magical Rabbis preaching magical lies; however, Cyrus the Great king of the Persian Empire, was a representative of his people, and into making a better world. So, Cyrus destroyed the country of Babylon, which is why the country of Babylon doesn't exist today, and he rebuilt King Solomon's temple in the year 539. B.C. for the humane Cohen tribe of the Hebrews to return and continue to teach the scientific way, to search for universal truth to create knowledge. However, the Hebrew Cohen tribe, wished to survive to preserve the flame of democracy for **future generations.**

When you have problems in Life, all you need is an accurate understanding of reality. Any other way, it will not create a solution. Unfortunately, today the Persians have religion, i.e., the illusion from an oppressive regime with a Muslim Babylonian religious leader, but not a more accurate understanding of reality. That is the only reason the Hebrews, and the Persians are enemies, because Hebrew science in the search of universal-truth is the enemy of Babylonian lying religious leaders.

The Jews, the Catholics, and the Muslims all have a magical man in the sky who love them, more than they love themselves. And it was created by Babylonian preachers' lies, that promote bind faith, foe leaders, for more money and privilege, **and kill and are still killing Hebrew free self-thinkers in the search of scientific reality.** The Muslims **built their preaching temple, over the King Solomon's temple of wisdom** so no new scientific knowledge will be created, and it is viciously guarded by ignorant obedient followers, to protect leaders privileges!

To put an accurate perspective on history, my family's ancestors have been reading and writing stories in books and debating in search of universal-truth to understand reality for over 3.500 years. In Europe, people have been reading and writing for less than 1,000 years from Hebrews' writing. Today their leaders want to keep European people, and the world ignorant in the magical illusion of knowledge, not in debate for more accurate understanding so that the hierarchy of greed for power and privileges can control the populace. It is happening in many parts of the world today because the gangs of the hierarchy (the regime) would rather die than work for a living, and it doesn't matter how much they destroy.

Understand this because it affects your life and all lives. The Greeks had the wisdom to create democracy. Yet, even in a democracy, the hierarchy (the regime) condemned **Socrates** to death or moved to the desert island of Djerba, in the Mediterranean Sea. All because the hierarchy lusted for more power and did not want to lose their traditions of believing they were important! In war, a soldier killed defenseless **Archimedes**, because there is no wisdom in war. He was a free self-thinker, not a follower, who discovered the earth went around the sun 1.400 years before Galileo, and he measured the earth. **He was wrong by 3.00 yards, from today's earth measurement. Because of religious hate and war,** the Persian and the Turk invasions killed all the Greek free self-thinkers.

Because the hierarchy lusted for more privileges, they did not escape to the island of Djerba and today in Greece, there is no flame of wisdom for the future generations to give to the world. That is why people must become free self-thinkers as a majority of the population in scientific debate, for the ability to find the wisdom to create representations of the people for the people, for a safer world. Today, the majority of the new generation is into understanding scientific reality!

As I was growing up in the mid-20th century, my grandfather and uncle were the representatives of wisdom for the Cohen tribe of the Hebrews in Djerba. The humane-Cohen tribe took care of King Solomon's Temple of Wisdom in the land of Israel **and built King Solomon's Temple of Wisdom on the island of Djerba**. Since this magical Mediterranean sandy island was relatively unknown, no destructive leaders came there to destroy the old books of wisdom to control the world.

Fortunately, the ancient stories of wisdom were safely guarded in the extensive library that my grandfather owned. **These were originally handwritten on animal skin in scrolls, and some of the books were thousands of years old.** Religious leaders would have destroyed these old books if they had known of their existence. For example, religious leaders destroyed most of Plato's precious wisdom and many other precious books of the past for the control of the lazy people through Babylonian scary cruel ways to control the people. **After the colonial French left, Djerba became a Muslim aggressive religion country.** The Babylonian Muslim religion leaders preaching lies for power over the country, the gentle Hebrew People of the Book moved to Israel to avoid the violence, and to safeguard the ancient books of wisdom in my grandfather's library, On the island of Djerba and in North Africa, all of the Hebrews who escaped to Israel, like my family, could not sell their property; therefore, they lost their property and possessions to the violence of religious liars inferior gang, to fill important. **So please don't tell me that religion makes people moral.**

When my parents took us to Israel, we were eight surviving children: four boys and four girls. As refugees, life for us was very hard, as it must have been for many refugees worldwide. However, I made the best effort I could to survive and help my family. My baby sister, Malka, was stolen at the age of one by a corrupt doctor and sold to another family for adoption. One of my brothers died when my mother had a miscarriage, he was fully-formed. He died from unsanitary swamps full of mosquitoes and a lack of food in Israel's refugee camps; and one brother was born in Israel. That left me with five brothers and three sisters. However, today there is no lack of sanitation, clean water, or food in Israel because the country was and is run by free self-thinkers creative humane-beings, Hebrews, in a democracy.

An exhilarating moment for me, my family, and the world, was when my uncle, Moses Cohen, arrived in Israel to convert the ancient handwritten books of wisdom into print and defend the democratic country. Even more so, Israel protects its people, so when I lived there in the 1950s, there were no homeless people in Israel and no homeless people still today.

When I was a lifeguard on the Mediterranean beaches of Israel, I worked 14 hours a day, seven days a week in the summer, and made more money than the President of the country, because this President was in it for the people, not the greed of money for more privileges. In Israel, 10% of the agricultural harvest of the land belongs to the people, therefore, anyone can walk in any agricultural land and eat all they want. This law was created by the Chon Hebrew people of the temple of wisdom to protect Democracy.

The people who founded modern-day Israel are free self-thinkers, Zionist Hebrews, not religious Jews. The Babylonian religious Jews did not want to defend the country; they expected their god magic man in the sky to do it. After 2,500 years, the Hebrews from Djerba's island moved back to Israel to receive the ownership of the key to King Solomon's Temple of Wisdom, where the seating arrangements are set up like a coffee shop, encouraging debate conversations to understand God's reality of science more accurately.

Suppose you are a believer and follower of illusions. You are a creator of hierarchy, which is a parasite, the enemy of all people, the enemy of a democratic world, and the enemy of the environment, from wanting more by creating inflation and being unwilling to give up their privileges, no matter how many lives and the environment they destroyed! For example. The God of the Romans, his name was Zeus, and his son was Jesus; he was betrayed by Juda in Jodea, by Jews, to make it easy to memorize the illusion of knowledge. That created a Babylonian magician's Jewish religion to compete to justify their existence. You can see how leaders write books and use words as hateful propaganda for the majority of the ignorant lazy believers to create blind-faith addiction, and to hate free self-thinkers. They must destroy

the temple of wisdom and preach lies to hate the free self-thinkers, and to control the world, by Atheist cruel selfish leaders.

Leon's uncle, Moses Cohen, of the hand-written and printed edition. My grandfather explained to me some of the images in the photo on the following page, my uncle has a religious Jewish hat, for Jewish hierarchy propaganda. For instance, the two hands with four fingers spread apart in the middle symbolize an upside-down pyramid. This image represents a "hand antenna" to collect wisdom particles from the Energy of the Universe. In the television show "Star Trek," Mr. Spock used that hand symbol meaning, "live long and prosper." The hand "antenna" helps connect with the energy and communicates with our inner selves because God is in us. It means ***Be in search of wisdom for a better world.***

My grandfather sayeth all free civilization Hebrews hath to pretend to be religious or they would have been extinct like other intelligent free self-thinking civilizations of the past because they are a minority. By preaching about the unknown's illusions to generate fear and encourage violence against all people who are not believers. Fanatics created by Babylonian magicians preaching propaganda about the unknown's illusion to create more privileges for parasite hierarchy by hate and war still today.

The first movement was King David's ignorant hierarchy, which believed God was on their side and would help them be victorious in wars. The result of that belief was that many people were killed or enslaved. Today, King David's followers are a religious Jewish movement because of the preaching about David's poems and songs that fervently praise and thank God for selecting David and his people as "the chosen ones, to feel important to be willing to die and kill for King David's hierarchy. That is stupid, that is the reason Christianity and Muslims copied King David's religious empire, to become a powerful parasite Babylonian hierarchy leader.

The second movement was understood by wisdom, the folly of war, as explained by John the Baptist, Jesus, King Solomon, and the Cohen representatives of the people. Those Hebrews understood they could not successfully defeat **the majority Roman army** *and their powerful hard-iron swords with their weaker bronze swords. They could not defeat the Roman army in battle, but a day will come when wisdom beats all the world's militaries from accurate understanding. That is why John the Baptist and Jesus said that King Solomon's wisdom inspired them; they came to conquer the mighty Roman army with their wisdom. In the end, they were victorious in history, and universal truth wisdom from accurate understanding will conquer all armies in the future.*

Today, 65.% of Israelis, are Hebrews, free self-thinkers, humane-beings, not religious in a tradition, as an example to the rest of the world, so more people will become free self-thinkers, humane-beings to create new knowledge to fix all the problems. The free self-thinkers who debate for a more accurate understanding question authority for a better life, because all religions create gangs of inferior people, and inferior people have to be bullies to feel superiorly important!

Leon's grandfather & uncle. Head of the Temple in Jerusalem.

Since the beginning of the Hebrews' movement, there were disagreements among the free self-thinkers who wanted to unite all people, including the religious ones. However, free self-thinkers, called Zionist humane-beings, created Israel's country by a scientific universal-truth, not by religious people in illusions! In the following picture, you can see my uncle, Moses Cohen, advising the brand-new Israeli government in the wisdom of the Hebrews in the humane law of democracy. They wear yarmulkes (hats) on their heads to remind them that the Energy of the

Universe God knows what they are thinking to remind them to be humane. However, you don't need to wear a hat if you don't need to remember to be humane.

דודי הרב משה הכהן זצ"ל מרביץ שיעור תלמוד לממשלת ישראל- 1958
(נשיא המדינה בן צבי, שר הדתות משה שפירא, שר המשפטים רוזן, מבקר המדינה נבנצאל ורב כץ מנהל בתי הבדק)

As the real story goes, a general named Dreyfus, a self-thinking Hebrew, was wrongly convicted as a traitor by a prejudiced French Christian court, just like the ancient courts did in judging Socrates. **A newsman named Theodor Herzl** observed the court case and understood there was no justice for independent thinkers. So, he established a charity called Karen-Kayemet, to raise money to purchase land in Palestine, Jordan. When I was ten years old, I volunteered to collect money for Karen-Kayemet, because they did not wont to still the land, and establishing a safe place for free self-thinking Hebrews. At first, they purchased all the swampland the Jordanians were eager to sell. They got it cheap because it was full of mosquitos, which thrived ever since the earlier Turkish invaders chopped down all the trees to be used as firewood to fuel military trains.

After the Zionists established themselves, they demanded their independence because they did not want their country run by ignorant of reality Muslim leaders that promoted hate, because they have nothing to offer. Religious Jews did not want to defend the country, because all leaders needed to protect their importance by their obedient followers.

The reason Zionists succeeded was that they were creative Hebrews, hardworking humane-beings using their free will in scientific reality. It did not derive from praying or showing their love for a religious leader because they are not gullible followers and that was the reason for their amazing success. After the Zionist success, religious leaders tried to credit the victory by saying God answered their prayers to promote their importance because God loved them more. God plays favorites. So why did God not answer their prayers about this new virus, Covid-19? Because more % of religious people are sick and dying than other groups globally. If you can justify this, can they justify anything?

The Hebrew Zionists did not wear funny hats or funny clothes; they were not, gullible believers in blind-faith. However, they understood the energy of the universe God and were successful free self-thinkers searching for wisdom to be creative, humane-beings, who did not expect a man-god in the sky to be responsible. They called themselves Hebrew free self-thinker Zionists, not Jews, because they did not want to be used in competing gangs for hierarchy propaganda to create hate and war to improve hierarchy privileges! Zionists are free self-thinkers and Moses' movement gives legitimacy; and the right to be a free self-thinker in this world is most important, because free self-thinking unites all the people for a safer democratic world and propaganda from believing in religion; and tradition, which separate people and creates hate and war by using the illusion of knowledge, to protect parasites' hierarchy, and that is against life, and everything that is against life is against God and against democracy. Today, the peaceful Muslims in Israel have a better life and better education than all other Muslims in the world. Eventually, the Israeli Muslim will become a free self-thinker humane-being in Moses' law, between people and God, with no religious parasite leaders (god-players), **as in the story of Jerico,** where the people united with the Hebrews; and that will create a powerful humane movement in a democracy, that unites all the people of the world!

You may recognize some of these famous free self-thinker Hebrew Zionists in the photo collage below. They are not followers of the hierarchy!

The Zionists knew they could not fight and win a war based on the illusion. So, the Hebrew free self-thinkers had to survive by uniting all the people, including the religious people to be tolerated by the world of dominating gangs created by the parasite hierarchy of the United Nations!

Many religious Israelites became Buddhists or atheists because they did not want to be part of their family's ridiculous carnival of Judaism. People don't know that if Israel's Zionists had not become friendly with all religions, they would not have received the help and support from all hierarchies to survive, because they are a minority. Now the free self-thinkers in a scientific debate have their country, representing the democratic government people, with astonishing success.

Their example shows free self-thinkers in a scientific debate, searching for universal truth for that time, representing the people in government works; and a hierarchy system that competes by memorizing knowledge that creates freedom of religion in a free self-thinking democratic government never worked before, and will never work, because religious leaders have nothing to offer. A utopian society is easily obtained by free will, to free self-think in a scientific reality method of debating, to arrive at universal truth for the time, and that will provide accurate understanding to understand what is right from what is wrong, and to create humane-beings out of people to do what is right, and that will eventually create a democratic utopian society!

On YouTube, a stand-in building or a book that does not give you the ability to disagree in a debate conversation are preaching propaganda, paid by wealthy Babylonian hierarchy, to stay in power **by creating scams, because it pays to advertise propaganda with your money**.

An example of what free self-thinking people gave you is Moses' democracy, the 10 Commandments, not Babylonian god-players that create religious lies between God and people. They got rid of all God-kings and hierarchy that steal your women and torture the people of their free will. They gave you Socrates, Archimedes, Jesus, Galileo, the Hebrew King Solomon who wrote the Masonic book to create the Masons who wrote the American Constitution "by the people, from the people, for the people" in the country of America. Albert Einstein, and King Qui of China who created the Wall of China, got rid of parasite leaders to create a great Chinese civilization and all the greatest humane minds by free self-thinkers. They are the ones that created and will generate more in the future. It was not created by Babylonian leaders that created religious lies that created sickness of blind-faith addiction.

Free self-thinking is an inspiring exciting joyful journey of challenges in search of universal truth. You can be one of them, and as a free self-thinker, you have the right to disagree with scientific debating knowledge, searching for a more accurate understanding of life's spectrum magic reality. Free self-thinkers never have torture chambers and never will; free self-thinkers don't have the death penalty; free self-thinkers have no crime rate.

They don't like illusory religious violent criminals, people who want money to do nothing, because words don’t hurt people, only truth hurts people, and pain is essential to tell you where the problem is. Imagine this as a free self-thinker to understand and in a debate find the truth, because swallowing many pain pills or believing in an illusionary religion, doesn’t solve the problem!

To make America great again is to go back to what made America great in the first place, a representative government in a democracy that gives power back to creative free self-thinking people, with no religion or leaders’ hierarchy, and that will solve all problems.

PART FIVE

Inspiring Photos and Life Stories of Leon

CHAPTER 39

LEON'S FAMOUS LIFE STORIES

I was born on the desert island of Djerba in the Mediterranean Sea, off the coast of Tunisia. Djerba was a French colony in the mid-twentieth century with a population of 4,000. There was no running water, no electricity, and no cars. All transportation on the island was provided by donkeys, camels, horses, bicycles, and feet. The residents of Djerba spoke many different languages, including French, Arabic, Italian, Spanish, Greek and Hebrew. When the ancient Romans ruled 2,000 years ago, they grew wheat as food for the Roman Army in Tunisia and lived on the island of Djerba.

One day, while I was on the way to my father's shop, I stopped in the center of town to watch a snake charmer's show, which started oddly. The charmer choked his cobra until it lost consciousness. *How can the snake perform now?* I wondered. Then the performer put small pieces of paper with religious lies written under the head of the snake. When the snake woke up, the snake charmer-preacher started preaching to explain how religious writing was so powerful that it caused the snake to come back to Life. He began selling pieces of paper for one franc each. When I returned to my father's shop, I confronted him and asked, "Why are grownups willing to pay one franc for a piece of paper?"

I was only five years old, but I knew it was just a piece of paper. Because I was taught to free self-think, and to search for universal truth reality by questioning. My father said, "The difference, my son, is their fathers terrify their children to believe and obey, and preachers by killing the free self-thinkers, terrify their fathers to believe and obey, so Babylonian leaders will control them. But I teach you free self-thinking in search of universal-truth of that time, so you can analyze everything for yourself. To better yourself and the world." My father also said, "When someone truly loves you, they will teach you to debate in the reality of science and knowledge for a more accurate understanding of the reality of Life, to become a humane-being. That creates healthy love and respect of life; this is God's universal truth.

I will not teach the illusions of knowledge that create obstacles in your brain, which only make you a believer that creates no confidence in **blind-faith addiction,** to allow leaders to control you. However, it's also true that if you debate together, you stay together as a loving free self-thinker, creative for a safer world. A Hebrew is not religious nor is he an Atheist or a gang or capitalist or communist or socialist, or a Jew. He or she is a humane-being, free self-thinker who loves and protects life-, for a creative joyful safer world. You are born a free self-thinker with common sense in the search of an accurate God's reality. However, the majority are brainwashed to believe lies in a magic show, as reality by Babylonian preaching, from the day you are born. Religions and Atheism as traditions in blind-faith create a gambling addiction, a sickness that is hard to cure!

Hopefully, these stories about my Life will give the reader a better understanding of the context from which this book is written, with great empathy, not sympathy. At seven years old, I had an army truck accident that paralyzed me from the waist down, at the end of World War II. I stayed at my grandparents' home for three years, where I learned about the reality of life. I had no feeling in my legs, and I could not move the wheelchair by myself. The technology to move wheelchairs on one's own did not exist yet. Therefore, my grandfather pushed me around; and, by asking questions and debating the reality of life, I understood the direct way necessary for universal-truth to love everything in life and be creative for a better world.

One day, a woman came over for some advice from my grandfather. She said, "The Devil is harassing me. Do you know what I can do to keep the Devil away?" My grandfather said, "The Devil comes from believing in (the illusion of knowledge) religion, not from an accurate understanding of reality. But if you walk down this street, there **is a religious place where they know all about the Devil."** Later that day, my grandfather said to me, "Grandson, things are changing very fast, and, in your lifetime, leaders will declare. 'There is no Hell,' because people are demanding a more accurate understanding of reality." **My grandfather predicted the future 65 years ago by his wisdom. In 2018, the Pope declared that there is no Hell.**

You cannot be creative, from illusion or Religions or Atheism or philosophy or tradition or by memorizing knowledge!

Only by understanding universal-truth, you can be creative!

One Picture is Worth More than A Thousand Words!

Don't just look at a photo, analyze it for yourself and see a thousand words. For example, my family is not hugging each other in the photo because they are not part of a gang, but rather free self-thinking individuals searching for the universal-truth to create a better world.

I was two years old, sitting in the middle of this family photo, full of curiosity and asking many questions while this picture was being taken.

All religions are fairytales created by Babylonian magician leaders (god-players)! All religious people have a Santa Claus in the sky, which loves them more than they love themselves. That is childish because in reality there is no Santa Claus in the sky or whoever you believe in. So please don't give weapons to children because they will play childish religious war games where they will hurt or kill themselves and others, and that is what people call stupid, so they believe that religion makes them moral, and that is the opposite, because believing in religion is uncertainty.

that creates criminals, creates Atheist, creates blind-fait, and that create lawyers........ **So please don't tell me that religion makes people moral!**

The man standing in the middle is the grandfather of the grandfather of my mother. He lived to be 129 years old. When people are humane-beings, they create the wisdom to have a long, peaceful, loving, happy, serene life. No doctors, no judges, no police are required. That is utopia. By limiting stress, our body and mind are capable of healing themselves. In the picture, my aunt standing on the right wears a hat full of gold coins. It is the custom in Djerba when a husband loves his wife that he gives her a gold coin which she happily stitches on her hat, so people can see how much her husband cares for her. You would not be able to practice this custom anywhere else in the world because somebody would steal the hat in a second! The women wear wooden shoes with a strap in the front, called Bakbak, to make a clapping sound when they walk. They have wisdom on the island of Djerba. When the men hear clapping, they know a woman is close by, so they take care and are respectful. My mother sits on the left and I am the two-year-old sitting in the middle. To this day, I still remember when this photo was taken. These people have not quarreled nor fought between themselves, nor even had a jail, or lawyers for over 2,500 years, because of their more accurate understanding of reality, which leads to a utopian life.

Violence is for animals—mainly to obtain food or avoid becoming food—but not for people who debate in a scientific method, which creates an accurate understanding.

For an accurate understanding of reality, I emigrated in 1966 to America, on my own. I knew nobody in the United States when I arrived, I could hardly speak English, as you shall see. Despite these obstacles, I earned a general contracting license and have since devoted my skills to creating inventions and innovative solutions for improving people's lives.

I invented the first sailing surfboard, aptly named the "Sail Surfer." I was awarded two patents for the board and was featured on the cover of Popular Mechanics Magazine. My sail-surfer design helped kick off the windsurfing craze of the 1970s. A few years later, based on decades of experience as an ocean beach lifeguard who'd completed over 275 certified rescues, I invented the first known standup "rescue paddleboard" in California. It was called the Multi Banana Boat, so named for its yellow color and slightly curved deck, which inspired today's paddleboards. The secret of its success is that a rescuer can slice through the waves to reach a swimmer in trouble, place him on board, create breathing C.P.R. and surf back to safety on the beach. I also invented the first therapeutic electrical stimulation necklace, named the "Solar Energizer" (patent N#.4.173.229—November 6, 1979), which was a big success, selling over 150,000 necklaces, which inspired today's electrical therapeutics devices and my rescue boat, (patent N# US 8,499,707 B2—July 13, 2005).

From a not-accurate understanding, people assume three realities into one understanding. The first reality is God's reality of science; by free self-thinking and debating in scientific conversation, you will find the wisdom to improve life for all lives. The second reality is imposed reality by hierarchy, to protect their privileges, and that is paper money, religions, traditions, propaganda, hierarchal laws, etc. The third reality is your own reality. This reality is subjective to each person. However, in a scientific debate, and by free self-thinking, you will create a more accurate understanding and realize that some of your reality is an illusion, opinion, or propaganda that will go nowhere and

create conflict. Therefore, you will accept that it is time to abandon your imposed hierarchal laws, religions, traditions, propaganda, and imposed reality and accept God's reality of science as the beast chois to evolve for the better, into a more realistic understanding of the magic reality of life, to become higher consciousness humane-beings to protect the environment and all lives.

If a believer knew, he would say I know; however, since they don't know, they say I believe. Believers are willing to kill and get killed over what they don't understand, and **that is VERY STUPID. Get over it! It is just an illusion created by parasite cruel Babylonian leaders who are continuously preaching hate for money because they memorize knowledge; however, they don't understand the creator in scientific reality! So, please unite in the universal truth of science to understand the reality of life, not in religion created by Babylonian leaders, that are viciously dangerous, and kill the disobedient, so leaders are safe!**

After the assassination of over 200 million people by the competing hierarchy in World War II, it was accepted as an understanding of the life norm. I was born during World War II, into a free self-thinking Hebrew Choen people of the book, in a utopian living!

My goal was not to populate or create more wealth or compete or have power over others. However, my goal was to be an inspiring example as a humane-being, for a better future for all lives, not to memorize knowledge for much more wealth by competing to become hierarchy. So, I can afford to overpopulate is not correct, and it is destructive. It is wrong. Because once you have a more accurate understanding of God's life reality, you will realize in wisdom, that every day and everything in life is beautiful, and life was meant to be lived by accurate understanding in a utopian humane world! The saying: "Absolute power corrupts," is wrong! All humane-beings throughout history with absolute power were not corrupt, because some were humane-beings representative of life, that protected free self-thinkers, and not just a being that protect is on interest!

CHAPTER 40

LEON'S INSPIRING PHOTOS

You can lay down and rest on the 25 lb. Halfon Rescue Boat.

Leon as a lifeguard in Kfar-Vitkin

Lifeguard Leon on a rescue mission in Malibu, California

Leon as a circus acrobat

Leon's lifeguard friends in Haifa, Israel.

The Halfon Rescue Boat, 25 pounds. Once on the water,
it becomes 55 pounds, making it very stable,
able to carry 600 pounds.
It keeps people safe from predators and chilly water.
It's also great for rescues in floods.

"Once a lifeguard, always a lifeguard."
Leon D. Halfon
Leon as a lifeguard in Haifa, Israel.

Leon was lifting classmates at the age of 11.

Leon demonstrates his Rescue Raft.

I love this place! Utah's Dirty Devil River.

Leon surfing his Multi Banana Boat in Malibu, California.
"If you are afraid to die, you don't live!"

Leon as a lifeguard in Nice, France.

Leon performing a Timex Watch Company commercial stunt.

Leon, the Timex stuntman in America's longest-lasting commercial —25 years on television, from confidence!

"I grabbed the baby like a football and I ran."

I was sitting at the lifeguard station and this lady came in with this baby. She went into the water, a wave came and knocked the baby out of her arms. Of course she started screaming, in a complete panic. I couldn't find the baby so I grabbed everyone and had them hold hands and feel around with their legs. After five minutes we found the baby. Of course she wanted her baby, but I had to give him some air. I grabbed the baby like a football and I ran into the room where I kept my surfboard. I closed the door, locked it from the inside and was working on the baby. And there was a crowd outside banging on the door, trying to break it down. Finally the baby regained consciousness. And as I handed him back to his mother, I realized that if I had not been able to save him, my God, they would have killed me.

LEON HALFO

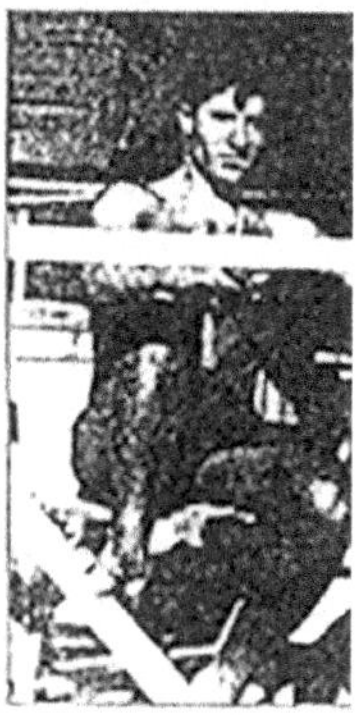

Silverlake resident Leon Halfon was a Haifa lifeguard before coming to California, where he developed and marketed a hybrid kayak/surfboard.

I was honored to be featured in the official book celebrating the 50th anniversary of Israel's formation, entitled An Eyewitness Account.

When, you understand, right from wrong, in God's magic reality of life, you will do what is right, even in the face of danger.

Acrobats Mimi and Leon are on a rescue mission.

Mimi navigates Leon's paddleboard, from confidence!

Acrobats Mimi and Leon on the Malibu Paddle Cat.

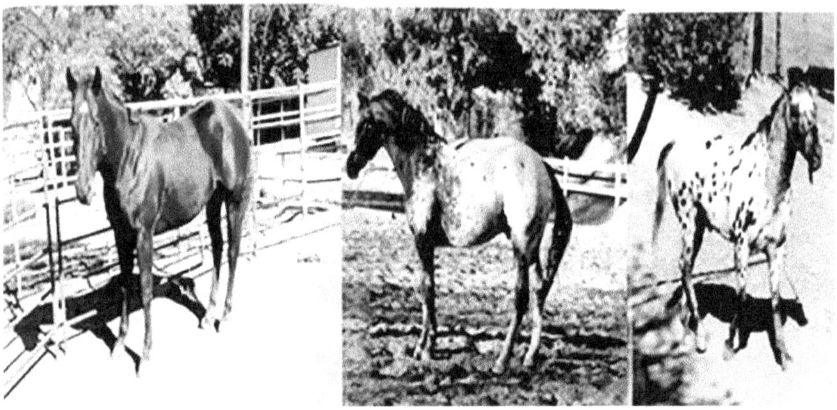

I love all living things—here are some of my wise equine friends.

Riding my horse, my goat follows me everywhere for the company, for the love of life, from confidence!

Leon feeds leopards at the breeding program in Florida for the Endangered Species Society, from confidence

Leopards and Leon.

My kitty cat is a dentist.

Leon with an animal friend.

At age 12, after recovering from my crippling injury, I could walk again. After that, I thought of myself as Batman, protecting vulnerable, free self-thinkers from bullies.

Leon, at the age of 50, realized that Life is beautiful and funny and if you don't understand that Life is lovely and funny, it is because you don't have an accurate understanding of the spectrum of life magic.

At the age of 60. My goal was always to stay in shape.

METRO

ARMANDO ARORIZO / Los A

Paddle in hand, Leon Halfon demonstrates his hybrid 72-pound fiberglass board along the surf of Santa Mon

Trying to Catch a Wave

■ Beaches: By demonstrating his floating contraption—part surfboard, part kayak and part sail board—inventor hopes to create a tide of attention.

By JOE MOZINGO
SPECIAL TO THE TIMES

What stops passersby is the odd vision of a stocky man strolling on the surface of the water.

The upright figure, wielding a 12-foot paddle through the reflected glare, moves through shoulder-high surf as he heads into an ocean glassed off by Santa Ana winds.

Witness Leon Halfon's offbeat sales method.

Every Sunday morning, he removes his hybrid of a surfboard, kayak and sail board from the chicken coop in his Silver Lake backyard and heads to Santa Monica to promote his dream.

A block south of Pico Boulevard, he unloads his invention from the rusted bed of his pickup truck and heads to sea. Sometimes he ambles north around the pier, or farther up the coast toward the reefs near Malibu.

"I go to the rocks and see the garibaldi and the seals," said Halfon, 55. "When you stand up, you see a lot more."

Onlookers squint to see him, as they undoubtedly did 90 years ago to watch Hawaiian George Freeth glide across the waves at Redondo. Billed as "the man who could walk on water," Freeth introduced surfing to California on his 200-pound solid wood board.

Halfon is trying to introduce his 72-pound fiberglass board—with an optional sail—in much the same way. But unlike Freeth, who was funded by the Los Angeles-Redondo Railway and industrialist Henry Huntington, Halfon works all week as a general contractor.

He has only three years of education but has invented th entire life. He came up with the this hybrid board while worki commercial fisherman in Israel. " was to slide fully loaded fishing ne floating board first instead of hau catch all the way up into the boat.

As a former lifeguard, Halfon the merits of a board as device that could be maneuvered the surf and used as a platform to CPR. But a county lifeguard who the board says it moves too sl rescues.

To get a surfboard through surf to go through a sometimes-jolting called duck-diving. Halfon just board, bends his knees and glows t

"When a big wave comes, it's Halfon said. "But I've crashed waves over my head."

If he's riding in and he wants t

Please see INVEN

On the first stand-up paddleboard for surfing, Leon in the Los Angeles Times newspaper on November 10, 1997.
| JOE MOZINGO | SPECIAL TO THE L.A. TIMES

Walking on top of the wave...
The most daring stunt, by the power of free self-thinking understanding of God's reality magic of life, from confidence!

CHAPTER 41

LEON'S FAVORITE ANIMAL STORIES

Falcon

One day, when I was ten years old, I walked by a grain mill building and noticed a dying baby falcon on the ground. I looked up and saw workers on the roof, so I yelled, "Watch out! You just knocked over and killed a baby falcon!" They told me to come up and take the rest of the falcon chicks because the building was scheduled to be demolished.

I took the tiny trembling birds home, but they would not eat when I tried to feed them. Searching for wisdom, I analyzed the situation. I realized that when they were in their nest on the roof, it was sunny up there and when the mother came to feed her chicks, she created shade and sound. So, I put the falcon babies in a box outside in a sunny place and then I took a piece of cardboard to create shade. They were chirping and hungry for food. I imitated the sound of the mother falcon and attempted to feed them. And guess what? It worked—by analyzing, and understanding reality in a scientific way, it would have not worked in believing or philosophy or illusion.

They grew up fast and were ready to fly in just two weeks. Every day when I came to feed the falcons, they would dive from the sky and land on my arm to eat. I would take the falcons to parties to make money as a show. After the falcons' performance, I would toss them in the air, and they'd fly home. One day one of the falcons did not show up, so I went looking for him. Not too far from home, on the roof of a four-story building, was the missing falcon. I recognized the lost raptor, so I raised my arm for the falcon to land and started calling and whistling. The people on the balcony watching me thought I was crazy because people in science seem to be crazy for people in the norm. The falcon dove off the roof at great speed and landed smoothly on my arm. Later that day, I decided to teach the falcon to hunt to be independent, but it was not easy. However, in God's reality of science fixing, everything is possible.

My family told me that the falcons would fly over to visit me whenever I left home and call for me if I wasn't there. I did it because I was a humane-being who had the wisdom to care for and love all Life, for the joy of life.

A YOUNG APPALOOSA STALLION AT MY RANCH

All my cowboy friends told me I had to break this beautiful wild horse's spirit and show him that I was the boss. Since I did not accept that kind of attitude, I thought to myself, *How can I train this wonderful animal without hurting him and his spirit?* I came up with a plan of wisdom.

I took the horse to a lake in a trailer, walked him into the water up to his chest and then climbed on top. He struggled a little bit but then got used to me and settled down. He was amiable after that! I could ride him without showing "Who's boss." When I fell off the horse, he did not spook and run away. Instead, he stayed and waited for me. That was big because it is essential for me to always be in friendship.

MY THOROUGHBRED STALLION

One day, I decided to breed my thoroughbred mare and searched for an appropriate stallion for her. Unfortunately, all the good ones were too expensive. Later, I went to a horse auction and heard a loud racket from an enclosed horse trailer outside.

I asked the owner, "What do you have there?" "A thoroughbred stallion," he replied. "However, he is a killer horse. He tried to kill his rider, and I am supposed to put him down. But I don't have the heart for it. So, what I am willing to do is haul him to your place for $175. And you keep the horse for free."

I never use aggressive behavior towards animals and people, after a while, this beautiful creature became my friend. I took him to the beach in Malibu so that we could ride together for exercise and fun. I built a corral for him along the Pacific Coast Highway close to Our Lady of Malibu Church, where I was employed so I could take care of him every day. The first time I rode Lush Life on the beach, we went into the ocean and a wave splashed us, filling my boots with water. I dismounted, removed my boots, and tied them on the back of the saddle. While riding again on the beach, one of the boots slid down and kicked the horse in the groin, making him run very fast. I couldn't stop him. I decided to ride him towards deeper sand to slow him down. Lush Life ran full speed into the deeper sand. Unable to gallop in the softer sand, he suddenly stopped, tossing me forward over his head. As a lifelong acrobat and instructor, I knew what to do. In mid-air, I executed a flawless somersault and landed on my feet. People watching from their beach houses applauded my graceful maneuver.

One time I rode my stallion in a Malibu fifty-mile endurance race. Out of 130 horses, my horse was the only one who was not an endurance racer. Even so, Lush Life and I finished number 8. The reason for our success was that the horse loved me so much. He was willing to go through hell because I asked him to. I cared for him very much and made sure not to push too hard. I checked his heartbeat, monitored his water intake, and dismounted to lighten his load so we could run up hills together. When he was nine years old, riding was too hard on his back, so I sent for his papers, which required the tattoo number inside his upper lip. (To be identified, all thoroughbreds have a tattoo number inside their upper lip!) That's when I first learned his name was Lush Life. I also found out that during his racehorse career, he won $180,000 in the Santa-Anita races. Had I known at first how valuable he was, I probably could have sold him for six figures. But I never would have because I loved that horse and wanted his years to be comfortable and joyful. And also, because I love all living things. So, I sold this handsome stud for $10,000 to the Malibu Horse Academy to breed and run free in pastures with mares. He was a great friend, and I miss him very much. Unfortunately, his previous owners did not accurately understand Lush Life or horses or the joy of life at all. They just wanted to win the race for more money instead of learning from each other and enjoying Life.

CHAPTER 42

LEON'S CALIFORNIA STORIES AND MORE

SWINGING ONTO THE BALCONY

At one time, I lived with a topless dancer named Beverly in a fourth-floor apartment, coincidentally near Beverly Hills. I had created a company that manufactured small fiberglass boats, kayaks, and canoes. One night, after delivering ten small boats, I was tired, very cold, and anxious to get under a warm blanket. But the apartment door was locked from the inside. So, I didn't know Beverly was there with another man.

At first, I thought of waiting, but I got colder. So, I knocked on the door, and rang the bell, but there was no response. So, I ran to my truck, grabbed an industrial-strength extension cord, and attached it to an air vent on the roof. Like a hero in an action movie—I'm thinking of Douglas Fairbanks in a 1920s pirate flick. I swung down from the roof and flew towards the balcony, knowing its glass doors had no lock. Before landing, I see Beverly and her boyfriend standing on the other side of the glass with their mouths open.

I land on the balcony and walk into an empty apartment. They ran away. I decided to sleep on the couch and unlock the front door just in case the lovers wanted to return and use the bedroom. The following day, I called the bar where she worked and asked for Beverly. I was told she was in the bathroom, so I waited and called a half-hour later again and the answer was the same—she was in the bathroom. I replied, "I don't want to play games. I'm coming down there to straighten this out with Beverly." When I arrived at the bar, it was closed, and I saw ten men milling outside. I asked what happened, and I was told, "The owner threw us out of the bar because he said a mighty man was coming over to destroy the place." However, I am a very gentle, humane-being who would always use conversation instead of violence to resolve an issue. Of course, if attacked, I would defend myself.

A few days later, I drove by the apartment, saw Beverly closing the balcony drapes, and realized she was there. I just wanted to talk and straighten out our differences. I ran up the stairs to our fourth-floor apartment and saw the door was open. When she saw me coming, she tried to close the door and lock it from the inside, but I jammed my foot to keep it open. I assured her that I meant no harm and told her that she could have it if she wanted the apartment. She said, "Okay." So, I gathered my stuff and left. That was the last time I saw Beverly.

I DATE A WOMAN WHO LIKES TO DRINK AT A BAR

When my girlfriend Lois and I stopped at a bar for a drink, I told her I wouldn't say I liked the place because it was too run down. She said, “No, no problem. Let’s just have a cold beer!” While drinking our beers, a drunk man walked over and unzipped the back of Lois' dress. She showed me what happened and asked me to do something about it. I zipped up her dress and told the man, “She’s with me. Leave her alone.” But the drunk came back and unzipped her dress again. I asked him, “What do you want?” He said, “I want to fight you.” I responded, “I don’t want to fight.” He ignored me and asked, “Do you want to fight in here or outside?” I repeated, “I don’t want to fight you.” However, to avoid a commotion in the bar, I said, “Let’s go outside.”

As we went outside, the drunk jumped on me and tried to sucker punch my face. I blocked his punch, hit him right in the mouth and knocked loose his two upper front teeth; they were dangling from his gums. He was spitting blood everywhere and looked horrible. Yet, he was laughing the entire time!

Meanwhile, I felt terrible. I realized I should’ve known better than this. I should have walked away. I didn’t sleep all night. The next day I went looking for that man at the bar to offer to fix his teeth. He wasn’t there, so I left my phone number, but I never heard from him again. Nothing seems to matter to the ignorance of the Creator or the drunk.

As a humane-being, even if you win a fight defending yourself or protecting others, you will feel miserable afterward because you are hurting someone. Therefore, the best thing to do is just to walk away. The prophet meant this when he said to turn the other cheek; he suggested walking away. Because a humane-being doesn't compete.

WORKING IN THE LITTON BUILDING FOR MAURICE BRASLAVE

Years ago, the Litton Building in Beverly Hills was a famously large construction with tall ceilings where airplanes were built. But after World War II, the company went under, so the building was subdivided into different sections and leased.

Maurice acquired the front area, which was about 5,000 square feet with a 30-foot ceiling above the concrete floor. His business in Beverly Hills, called "The Last Show," produced exclusive one-of-a-kind hand-crafted clothes and costumes for over 5,000 television and motion pictures, such as Marlon Brando's *Last Tango in Paris*. He also designed apparel for famous political and show business luminaries, such as President John F. Kennedy, California Governor Jerry Brown, and former Russian President Nikita Sergeyevich Khrushchev. His clothes cost 50 times the cost of regular clothes; however, his clothes made you look like a god.

One day, Maurice called me and said the fluorescent light he needed for his job was not working. I came over without a ladder and climbed up the sprinkler system's pipe, like Spider-Man scaling a building. I replaced the fluorescent light, but it still did not work. I climbed down, walked to the circuit breaker box, and switched the lights off. Then I flipped off the circuit breaker to the fluorescent light and taped the switch so no one would turn it on while I was working. I climbed the pipe again, disconnected the ballasts from the lights and left to buy some new ones.

When I returned to replace the ballasts, I climbed the narrow sprinkler pipe to work on the light once more. Somebody from another part of the building came to the circuit breaker box, took the tape off, and switched on the lights. The building used 280 volts of power. Suddenly I felt pain throughout my chest and heart. I was being electrocuted from sitting on a cast iron sprinkler system pipe. I was almost paralyzed and could not let go of the wire, so I tilted myself to let go and fell to the concrete floor 30 feet below.

The ambulance took me to the hospital and x-rayed my back. The doctor said they would have to operate immediately since I cracked a bone in my vertebrae. I told him that I did not want anybody operating on my back and wanted an ambulance to take me home. I said, "I'm a physical therapist and I'll get myself better."

The doctor said, "You will have lots of pain and eventually will have to see me again. By then, it will be tough to help you. You will probably be in braces for the rest of your life!" I boasted: "I will go home and get myself better and then I will come back here and kick your butt." The doctor said, "I'd like to see that. I'll be waiting for you." Sadly, back in those days, back surgery was not often successful. I went home, built a cast out of flour, eggs, water, and cloth, and wrapped it around my body. My father was a medicine man on the island of Djerba. I stayed in bed for six months with light exercise. My friend Cosmo took care of me during this time. After some more challenging workouts, I was able to get out of bed. It took me a year before I could leave the house. Two years later, I went back to the hospital and told the doctor who wanted to operate on my back to bend over because "I'm going to kick your butt!"

Here, the lesson is that you must free self-think wisdom for an accurate understanding that creates confidence and trust in what you understand in reality, and everything is possible. By free self-thinking, you will see the truth of Life. Work hard and anything is possible.

THE VOICE OF EUROPE

In 1966, when I first arrived in California, Robert Bardey and Gerard Alcon became my new friends and I joined their work as reporters for a Paris radio station called *The Voice of Europe*. We recorded news stories and events on tape for almost three years and then called Paris on a transcontinental landline.

To broadcast on the air, we put the telephone on a recorder and instantly transmitted our information to Parisian radio listeners.

THE PREACHERS

One of the events we recorded for the Voice of Europe took place at a park in Long Beach, California, where preachers use soapboxes. After spotting a group of people, they would stand on their soapboxes and start preaching. A preacher in front of us walked like he was drunk. We followed him and when he arrived at the crowd, he stood on his soapbox and started preaching. While we recorded the event on tape, the preacher confessed that he used to be an alcoholic, but Jesus saved him, and he was no longer an alcoholic. Then Gerard approached the preacher's soapbox, pulled him down, smelled his breath, and told him, "You are still an alcoholic. Phew, you stink! And you don't deserve a soapbox." Everyone in the crowd started laughing. **That is a good lesson for you, about Babylonian preachers**!

THE MUSTANG RANCHES

One time, we went to Las Vegas, Nevada, to interview sex workers in the now-famous brothel, The Mustang Ranch, for a radio documentary. When we entered the lobby, twenty young women were wearing bikinis and standing at a long table. An old woman would say to the visitors, "Hurry up because the young ladies are getting cold." Once a visitor selected a woman, we followed them to their room to record the event. She gave the man a pan with soap and water, and he wondered why. He asked, "What's that for?" She answered, "This is what we call a 'peter pan.'" He said, "I thought Peter Pan was a flying kid." She replied, "Peter Pan is a flying kid in a fairytale, but this is real: it's a 'pan' to clean your 'peter,' stupid."

MY CANCER STORIES

I had been with my doctor for more than ten years. Sitting in the waiting room for the doctor, a woman that I didn't know entered the room and said to me, "Why don't you act your age? You are 68 years old, and you had a heart attack. Who do you think you are?" I replied, "I never had a heart attack, and I don't understand your problem. I don't know who you are; I want to see my doctor." Later on, my doctor says to me from out of the blue and without any proof, "Leon, you have cancer. You always knew one day that was going to happen." No, I never assumed one day I would contract that deadly disease. However, I had to go for an MRI, undergo weekly blood tests and take numerous strong medications because of what he said. As a result, I fainted, fell, and severely broke my right shoulder.

Two weeks later, I ended up in the hospital with a fever so high—106 degrees—that the doctors thought I would not recover and asked for my permission to stop treating me and let me die. It took a millisecond for me to respond, "Hell no! I want to live. Do everything you can to keep me alive." Finally, another doctor in the hospital came to my rescue, treated my situation, and told me that the medication the other doctor prescribed was what made me sick in the first place. So, I got rid of my old doctors and the lousy medicine that almost killed me. After thorough testing, I found a new physician who concluded that I had no cancer and never had cancer.

I made an appointment with my original doctor, who had affirmatively told me earlier that I had cancer. I confronted him and asked why he gave a false diagnosis. The doctor looked up his computer records and said with a straight face, "I never said you had cancer. "Meanwhile, all this time, I was worried sick over nothing. The doctor's untruth was very destructive to my mental and physical health. What this incident showed is that some psychopathic people in the hierarchy play God because they can. What a relief it is to know the actual reality, so keep on questioning. In wisdom, believe nothing of what you hear and half of what you see. People with knowledge and no understanding of God's universal-truth are dangerously destructive. Today, I am 82 and feel heathy. **So, please, join the free self-thinkers in the search for universal truth.**

CHAPTER 43

LIVING ON THE DESERT ISLAND OF DJERBA

My mother, Rachel, came from the Hebrew Cohen humane people on the desert island of Djerba. They were descendants of Moses' family. They didn't live by violence; they lived by debate, to create understanding. Before getting married, my mother was a schoolteacher who taught the French language. She was also fluent in Arabic and Hebrew.

My father's name was Mero. He came from a pirate family in Tripoli mariners lost tribe of the Hebrews, the Zebulun tribe. He spoke Italian, Arabic, Hebrew, and French. He was powerful and never backed down from a fight. He enjoyed roughing-up fanatic, religious bullies who go around beating up "non-believers." The French police appreciated what he did. He risked his Life every time he fought. But mostly, he was very friendly with all the citizens of Djerba.

After work, my father would invite Muslim, French, and Greek friends to drink sap from palm trees called Logmy. When fermented, the juice transforms into a deliciously sweet drink that makes everybody drunk. I was just a kid but sometimes participated in drinking it too.

Everyone in town knew me and respected my father. He taught me to think and analyze everything and not believe in fairy tales, Santa Claus, religious doctrines, or even birthday parties. We never had birthday parties growing up. When you reexamine your experiences, it creates a powerful pattern in your brain. If you analyze everything again later in Life, you'll remember things very clearly. From the age of two, I would ask questions. I still remember things as if they happened yesterday.

My father awakened everybody at six in the morning for a swim, in the summer. Once we were at the beach, he threw a piece of wood into the Mediterranean Sea and told us that whoever swam out and brought the wood back would get the treat of a pirate breakfast. But of course, there was never enough food because we lived on the desert island, and we were such a large family. So, I was always hungry and eager to claim that piece of wood.

I didn't know how to swim on the surface, so I held my breath and paddled underwater until I spotted the wood above. Then I came up, grabbed it, took a breath, and swam back underwater; the pirate breakfast consisted of a raw egg from which the white had been removed and replaced with half an inch of Arrack, which is similar to vodka and half an inch of olive oil on top. Whenever I think about those pirate breakfasts, I become a little bit nauseated, but when you're a hungry child, everything tastes good. So that was breakfast, the pirate's way. I was two years old at the time. When I was eighteen and became a lifeguard, I could hold my breath underwater for over five minutes!

When I was four years old, I walked to school with a Greek boy named Garibaldi. On the way, he would climb trees to eat raw birds' eggs and even baby birds to show that he was a challenging Greek. I particularly remember two rows of mimosa trees on the way to school. They smelled wonderful. I liked to pick the trees' yellow flower balls and sniff them, turning my nose powdery yellow, and making us laugh.

My best friend was Jacque, the school principal's son, who lived on the school grounds. On the first day of school, my three-year-old sister peed in the classroom. The female teacher took her to the front of the class, picked her up, and slapped her bottom. I was five years old and did not like seeing my sister getting spanked, so I ran up to the teacher and started kicking her. The teacher was jumping and laughing at me. Then, she told me that I was adorable and brave for defending my little sister. Afterward, she gave me many presents, including a kitten.

Another day I was sitting on the sand in the school backyard and a big beautiful African eagle landed on top of a 10-foot palm tree next to me. I stood up and saw that the eagle was standing and pacing, looking at the littlest kids as if they were to become a meal. Sure enough, the principal appeared with a rifle and shot the eagle down. That might seem wrong today—but even if a man threatened the children, the principal would have shot him. That was Life on Djerba.

After school, I would walk to my father's marble and wood carving workshop to help out, or attend Hebrew education classes, or play with the animals in our backyard. When the native people came to town to sell or buy things, they tied up their animals in the yard behind my father's shop. As a result, there were always more than fifty new animals to see every day.

In the backyard, there was a mother goat with two goat kids. My father let me keep the little male. I fed and trained him to understand a few words. He was beautiful, with big green eyes, pink lips, and rapidly growing horns. To protect me, the goat stuck by my side like glue. Maybe he understood I was a "kid" too. If someone came too close to us, he'd stand up on two feet, eager to butt the intruder away. That's the reason I didn't need a leash to keep him close by. I also taught him how to butt on command.

On the weekends when I was five years old, my goat and I would walk to the schoolyard to meet my friend Jack and his little dog. Together we created a small animal circus to entertain our friends and fellow students who came to watch. It was great fun for all of us.

One day my father told me it was time for us to eat my little buck goat! I was shocked. I couldn't understand why anyone would kill and devour such an intelligent, beautiful animal. I told my father that my buck didn't do anything wrong and was brilliant and friendly. I cried and cried so much that my father ended up selling him. Ever since then, I have not eaten red meat.

Sometimes when I didn't have to go to school, I went with my father in the morning, to the town square, which the ancient Romans built. They had constructed a concrete platform, four feet high, fifty feet long, and twenty feet wide so that the Djerba fishermen could display their catch for auction. Shoppers came every day early morning to buy fresh seafood because there was no refrigeration or ice on the island; fish were attached to strings in boxes for auction **and giant sea turtles were on their backs,** struggling before being sold for food. When the auction started, my father would bid on a string of fish. After he bought the line, I picked up the fish and walked home while my father went to his workshop. At the end of the auction, the rest of the fish were given to the poor.

After school, when walking to my father's shop, I felt like I was the only kid who was curious and questioned things. Because I was confident in myself, I wanted to learn as much as I could. From the age of five to seven, everybody was friendly to me, and I felt loved by all. Maybe that's why I love people and all lives to this day. People nicknamed me Wildcat because I was like a feline walking around everywhere. One day, I saw this older man having his afternoon coffee. He held a tail from a cow for swatting flies. One of the flies bounced onto the cup and fell into his coffee. The older man picked the fly out with his fingers and brought it to his lips. He sucked the fly then threw it away. I asked him why he did that. He said, "The fly adds extra flavor to my coffee. Try it sometime." Thanks, but no thanks.

One day walking on the way to my father's shop, I saw a man hanging upside down by his feet in a storage building with other people standing around. I came to look and saw blood dripping from his nose into a pan, so I asked the people what they were doing to this man. They said, "We are taking the Devil out of him." I did not understand at the time because my family never mentioned the Devil, but now I realize this was an exorcism, Muslim style. Sometimes, musicians and performers came from central and south Africa to entertain us in our town center. There was always something going on in Djerba. Often, Muslims behaved violently over religious matters. For the fanatics, anybody who was not Muslim was an infidel, not a follower of their religion in Arabic. In such a society, free self-thinking in wisdom was not permitted or practiced. It was punished, which empowered religious leaders to keep people childishly ignorant and obedient. How convenient for the leaders! That is what happens when there is no separation between hierarchic religious gangs and the government. They are the same.

Sometimes to defend himself, my father had to fight these fanatics. He said he didn't want to fight and would ask them to stop, but if he must fight—he'd fight to kill. I saw my father kill three men in self-defense. What a waste of precious Life. Bullies realize they are ignorant from believing in illusion, which explains their aggressiveness. These religious hellions become aggressive to feel sure about themselves, with their preacher's or leaders' encouragement. How convenient for the bullies caused by leaders preaching the illusions of knowledge that create ignorance of reality, they felt that their **illusionary religious god** would help them triumph and they would go to Heaven, just as their religious leaders had told them. You would think people would know better, but they do not, because they have a blockage in the brain, from blind-faith that makes them unable to analyze in reality!

I visited the island of Djerba in 2008. I saw a modern airport, automobiles, running water, electric power and five-star hotels. Tourists were everywhere, but at a cost! The sea was polluted; the only catch was sand crabs and four-inch flounders. The Roman ruins were gone, and the mimosa trees were missing, replaced by fancy hotels and trendy shops. It is an example of what is happening to the rest of the world: over-development.

My friends and I went to buy a Berber carpet. The Berbers were the only people who conquered the Roman Army. The Berber leader's name was Anibal. However, when the Berbers came home to Tunisia, they were betrayed by the Tunisian leader's greed for Roman money. The Tunisian government forced the Berbers into slavery. They are still slaves of the Tunisian government today, making Berber carpets.

When the police were not looking, my friend, Vesta sneaked $100 to a slave. I wanted to give money too, but I would have been caught by the factory manager who watched us. I gave $50 to share in the gift. Neither of us would buy anything from the carpet factory because we did not approve of slavery. Berber slaves have an H-shaped green tattoo on their foreheads, marking them as slaves of the Tunisian government, even to this day!

I tried to have friendly conversations with the native Muslim people in Arabic, telling them about my past on the island. Suddenly they surrounded me, demanding to see my passport. They wanted to know if I was a Muslim or not and if I left before or after the French colonial occupier's revolution. They were a gang of inferior bullies, so I got out of that area fast. What I learned is that in a situation like that, people often get stabbed. If I got into a fight to defend myself, I would not have had a chance to be found innocent of self-defense against a prejudiced Muslim religious judge because religious law is not humane law—it is a discriminatory law. So, all religious laws are prejudiced against non-believers, **they just protect selfish leaders.** Currently, Djerba is becoming worse, and so is the world, not better. In these Muslim countries, there are no representatives of the people, for the people. Religious leaders ruin and control the government and all of Life. Consequently, these countries are in a state of total unrest, which will not stop until they have humane laws to protect all individuals. The greater the religious illusions of knowledge, the greater the violence from ignorant bullies

CRUSHED

A few years after World War II, in 1950, a fleet of French Army trucks landed on Djerba's desert island when I was seven years old. The streets were narrow and not designed for motor vehicles. I walked next to an ancient Roman wall from school to my father's workshop when a French Army truck slammed into me and crushed me up against the wall. To call them "injuries" does not adequately describe the damage to my body. My spine and pelvis were smashed and there were broken bones everywhere. My child's body was no match for 5,000 pounds of careening military steel. Awakening into the world again, there was a presence in the room after two weeks in a coma. The silence was broken with a voice that sounded like mine: "What am I doing in the hospital?" My mother answered: "You had a terrible accident."

Struggling through the fog to understand what had happened, words echoed from my mouth, saying, "We were together when we went to see my brother after the truck accident. My father was beating some man. Everyone was screaming, 'He's just the insurance man!'" I saw people carrying his body. Then my mother interjected, "You could not have seen anything like that recently. You were in a coma. Everybody thought you were dead." So, the vision during my coma was impossible. I had an out-of-body experience, or I was in my brother's body. *I don't know, I never found out and I could not find out how that magic of reality happened.*

After regaining consciousness, I found myself bound in a full-body cast. The only body parts out of the form were my head, fingers and two small openings to pee and poop. The doctors believed I broke every bone in my body even though there were no x-ray machines at that time on the island of Djerba. The doctors sent me to a hospital in the city of Sfax in Tunisia. Because of the swelling of my brain, they drilled a hole in my skull, which they filled up with powder named sulfa-med. The hole eventually healed, but it left a bald spot on the back of my head the size of a nickel which didn't disappear until I was 40 years old. I stayed in the hospital for six months because I could not relieve my bladder after the accident. Everybody thought I was going to die unless I urinated. My bladder had to be drained every day using a syringe inserted in my penis, which was very painful.

A group of very loving Muslim children visited me every day, bringing candy and cheering me on to pee because they knew if I didn't, I would die. When the doctor drained my bladder, it was not as painful as when the nurse did it. One day the nurse came in to insert the searing into my penis. I screamed in pain from the searing. Ali, one of the Muslim boys, said, "We'll pick you up and stand you on the table so that it will be easier." As Ali and his friends did so, they chanted to encourage me to pee while the nurse tried to insert the searing. Suddenly, I started pissing on everything—including the nurse. Everybody cheered, except the nurse (even though she was happy for me). Shortly afterward, I was well enough to be released from the hospital, so they put me in a wheelchair and took me to my grandparents' house in a horse-drawn cart.

However, I still had no feeling in my legs, and I could not move the wheelchair by myself. The technology to move wheelchairs by the rider did not exist yet. So, someone had to push me around. Every day I was wheeled across the street so that I could get out of the house. If I wanted to drink or eat something, I had to wait until the evening to be wheeled back home.

I asked everybody who walked past me if they could run. They said they could, so I asked them why they didn't run very often. They told me they didn't want to. I could not imagine not wanting to run, when one could do it. So, if I could run, I thought I would run all the time.

Children were playing around one day and one of them hit me on the back of my neck because he knew I couldn't defend myself. Later, I declared to my grandfather: "There is no God. I would never hit someone for no reason, yet the kids hit me and ran away laughing. Where is God?" My grandpa replied, "God is the biggest and most powerful force in the Universe. You should be thankful for what God gave you." I replied, "Well, I am not thankful. " I didn't accept my grandfather's answer as truth because you will know whether a statement is a fact or fiction if you think for yourself. As a free self-thinker in wisdom, it is your right to disagree.

When I was outside in the wheelchair, a young French girl named Nicole came every day after school and protected me from the bad kids. I asked her, "Why are you so nice when so many other people are not?" She said, "I am a Christian, and Christians are supposed to be nice." I smiled and said, "I would like to be a Christian." My response indicates that I did not fully understand the reality at that age. However, it is wiser to be humane.

Another day, bad kids sneaked up behind me. They hit and harassed me and then ran off. I was sitting alone in the wheelchair, feeling miserable, when an older man showed up. One second, he was on my right side—the next second, he was on my left. He seemed magical. I asked, "What do you want?" He said, "Hey, kid! Why are you crying?" I answered, "The bad kids hit me again. And there is no God!" He responded, "Do you know what life all is about?" I said, "No, I am only seven years old. How should I know what Life all is about?" "I'll tell you," he said. "Life is nothing but a continuous journey of exciting challenges. God must love you very much since he gave you this tough challenge. Overcoming paralysis is like climbing the tallest mountain. Once you climb the tallest, all the other mountains seem like little hills. Do you want to find God? Accept Life as a challenge! What do you have to lose? Everybody is going to die anyway!"

After he left, I started thinking and analyzing what he said, just as my father told me to do. The older man was not trying to make me believe anything and he didn't promise me a rose garden, only challenging work. This I could do. *What did I have to lose?* I thought of myself as a free self-thinker. This seemed like the truth to me! This fact gave me the will and determination to overcome my challenges. So, I decided to build up my upper body strength since my lower body was useless. I started exercising by pushing myself up and down on the wheelchair's armrests, and once I felt strong enough I raised myself into a full handstand on the wheelchair's armrests. One day I felt strong enough to drop to the floor, raise myself upside down into a full handstand, and literally "walk" across the street on my hands toward my grandparents' house to have a drink or eat something.

This was my first realization that challenging Life is rewarding and exciting and it was a beautiful rose garden! Even with thorns.

Another time I was sitting in my wheelchair when a kid came over and tried to hit me. I jumped out of my chair, stood on my hands, and started chasing him upside down in a handstand. He ran away, scared, calling for his mother. I walked back on my hands, climbed into the wheelchair, and raised my arms over my head in victory.

I felt great. At that moment, I understood there is an energy in you that we don't understand, but it is accurate. That was the moment I was ready to challenge myself to walk. But until I could walk on my own two feet, I continued to get around on my hands with my legs dangling above.

Due to nerve damage, my legs were numb. Sometimes I would poke my legs with a needle so that I could feel something. After this, I worked incredibly hard to challenge my body to walk again. In the end, it was not a belief or a miracle from God that allowed me to succeed, because nobody is born important; you are what you do, not what you believe you are. It was my accurate understanding of the truth of the reality of life in free-self-thinking, by free will. Life is a journey of challenges. I met this challenge by putting forth the effort to exercise. The result was that my body healed itself because, God, the energy of the Universe is in us and around all of us.

Eventually, my father made my crutches and gave me ostrich oil to rub on my atrophied legs, which had become extremely thin and fragile from lack of use. After that, I got my entire body working again by exercising practically all day, every day, and I used needles heated over a candle to jab my legs to see if I could feel anything. Eventually, I began to feel something, which gave me hope of regaining the entire movement one day.

It took me two years to overcome my shattered pelvis, broken femur, and injured spine to walk again. My left leg is still 1 and ¾" shorter than my right, but it doesn't bother me. I manage very well now, even better than most people because I accept Life as a challenge and I did it my way by using free will, not blind-faith.

After intense exercise, I would fall into a trance and daydream in my chair. One day, I saw a storm approaching. I was not dreaming, but I felt frozen. The sky was cloudy and becoming darker and darker. The clouds seemed to be alive, pushing each other in a struggle. The clouds crashed into each other and created sparks of light. I was helpless and became very scared. I was frozen by paralyzing fear. I couldn't get up to leave or even cry for help. I looked to the clouds for an answer and saw a tiny red glow, about the size of a pea, in the corner of one of the clouds; I understood from my grandfather that God was Energy, so I concentrated on just looking at the light for help and not paying attention to anything else. The energetic light got bigger and bigger. The light came down around me like a veil and I became engulfed in it; it felt warm and comfortable, but I could not comprehend it. Then, as if my body did not exist, I began to communicate **telepathically** with the light in an accurate understanding of meaning, not by word-conversation. I asked one question: What happens to me after I die? I began to understand the light's response: Everything in existence changes form but will always exist and always has existed. I felt as if I was communicating with myself.

YOU CANNOT MAKE SOMETHING INTO NOTHING,
AND NOTHING INTO SOMETHING,
IF YOU KNOW YOU EXIST IN YOUR MIND, THEN YOU ALWAYS HAVE,
AND YOU ALWAYS WILL EXIST.

When I woke up, I felt like I had just time-traveled. I understood the wisdom of Life at that moment. I was only eight years old, and my understanding was that I didn't need to procreate because too many people destroy the environment. We are the only species on the planet capable of making a better world, my goal in life. However, Nature needs to procreate to evolve better and more beautiful species. Life is a continuous cycle of birth, death, and rebirth.

People believed that if they walked to the edge of the Earth, they would fall off, but that's not true. Others think that when they come to the edge of their lives and die, they too will fall off, out of existence. That also is untrue because everything in the Universe always exists and only changes form.

If you can understand the meaning in your mind, of the paragraph above, you can better understand the infinite. Acknowledging this left me without a fear of dying, but of course, I prefer to live. Fear of dying holds people back from living a full life. Now I felt I could understand Einstein's Theories of Relativity by a scientific method of debating and free-self-thinking, which creates a more accurate understanding to create a healthy mind and body.

I continued my exercises, ever more determined to walk. It was absolute torture. I would go into a trance when exercising hard and wake up when it was all over. As a free self-thinker, I understood that people and animals could go into a trance-like state, so they will not feel pain because the Energy of the Universe will not have it any other way. In the beginning, I used crutches, but I barely put my legs on the ground, mainly utilizing my upper body strength to move. I slowly put more and more pressure to strengthen my legs until I didn't need crutches anymore. Finally, I could walk again! I felt victorious. Then I started running again. I realized the doctor and my parents didn't think I could do it. They expected I would stay in my wheelchair forever because they did not understand the power of free will. But I knew better. When you gain a more accurate understanding of reality, you will realize how to reach your full potential. My free-will reaction was that my body healed itself because the Energy of the Universe helped me understand how to reach my full potential. After this achievement, I was stronger and faster than I had ever been. I felt I fully deserved to live a life that suited me best. However, I did not always think this way.

As I mentioned earlier, my grandfather had an ancient library where I learned Hebrew. He had a book that I loved to read called *The Wisdom of King Solomon*. This is where I first learned about Masonic wisdom. I became an optimist as I began to understand the importance of the infinite wisdom in Life. I heard that America was based on Masonic wisdom, so I always wanted to go to the United States.

I was born at home within a tradition. I fought my family about having a Bar Mitzvah because it was a branding process like someone would use to mark animals or slaves to be controlled. My understanding came from ancient Hebrews' wisdom to unite all people worldwide by becoming a free self-thinking humane-being, not from religion. Traditions derive from dead leaders of the past to create and control "gangs." However, it is time to unite people by achieving a more accurate understanding of reality to arrive at universal truths that will make parasite hierarchy absolute, by creating a majority of the world, as free self-thinkers in a democracy.

I loved experiencing Life very much, but I was not a believer and didn't obey very much because I enjoyed my freedom. I was away from home most of the time. One day my father said to me, "You better obey! I brought you into this world and I can take you out of it." I said, "You'll have to catch me first!" I understood the importance of free self-thinking and self-preservation. Early one morning, while I was sleeping next to an open window, my father woke me up and slapped me. He said, "Who do you think you are?" I quickly jumped out of the window and ran two hundred yards to the Mediterranean Sea. The waves were big from last night's heavy rain. I dove into the sea and swam like a fish. My father got scared and started screaming, "Come back, Levi! I will let you do what you want (Levi is my birth name; my family called me Levy and everyone else called me Lee, which later evolved to Leon when I came to America). Often my parents were exasperated by me because I was highly active and challenging to handle. Everything became more relaxed after that. It is a good thing my father was not religious. If he were a religious follower, I would have been treated as a nonbeliever and burned alive at the stake. *Can you imagine that?*

So, stop believing in Babylonian magicians that created religion and tradition in blind-faith to keep you idle from illusionary knowledge lies, and start Free self-thinking in God in reality of science for a continuous journey of exciting challenges to understand and live your exciting life journey of challenges, and to do what is right, even in the face of torture, for a better world for all!

Use your mind, not your brain, because the brain as obstacle from believing in blind-faith of tradition. I hope you remember all the scrimping between my father and his brother about my accident money. No matter, I always loved and respected my father and mother, and still do; however, as a free self-thinker, I cannot do wrong, when I understand what is right in God's reality, even in torture, for a better world, for life.

My father told me he had 1 million and ½ francs from the insurance that belonged to me. All insurance paid 10 times my medical expenses. As a Lifeguard, I made 1,800,00 Liras per month, and my father made 400,00 Liras per month. Do I have to obey religions, or traditions for my family to treat me honestly? As a free self-thinker, I gave you a dog, falcons, a boat, America, and a horse ranch, to understand the joyful love of all lives. A laggard job is not a temporary job; lifeguards, at 80, are still lifeguards!

Me and Susy came to the house, and asked my father to marry us, when my father found out she had money, he asked me to give Susy to Simon, because he is my older brother. Susy and I were not interested.

CHAPTER 44

LEON THE LIFEGUARD

Being an ocean lifeguard was the best job of my life. What wasn't there to love? I loved the beach, loved the sun, loved fun, loved swimming, loved saving lives, and loved the pretty girls particularly and they loved me.

When I went to Israel, I was not educated because I had missed school from the truck accident. Also, when I was in school, I was bored. I needed a job, so I participated in the challenge to be a lifeguard. Unfortunately, out of one hundred and thirty-five people, only seven passed the test to be a Mediterranean beach lifeguard. However, I knew I would pass the test because I was a great swimmer, and I could hold my breath underwater for more than five minutes. After all, I was an islander from Djerba.

Woman losing bra

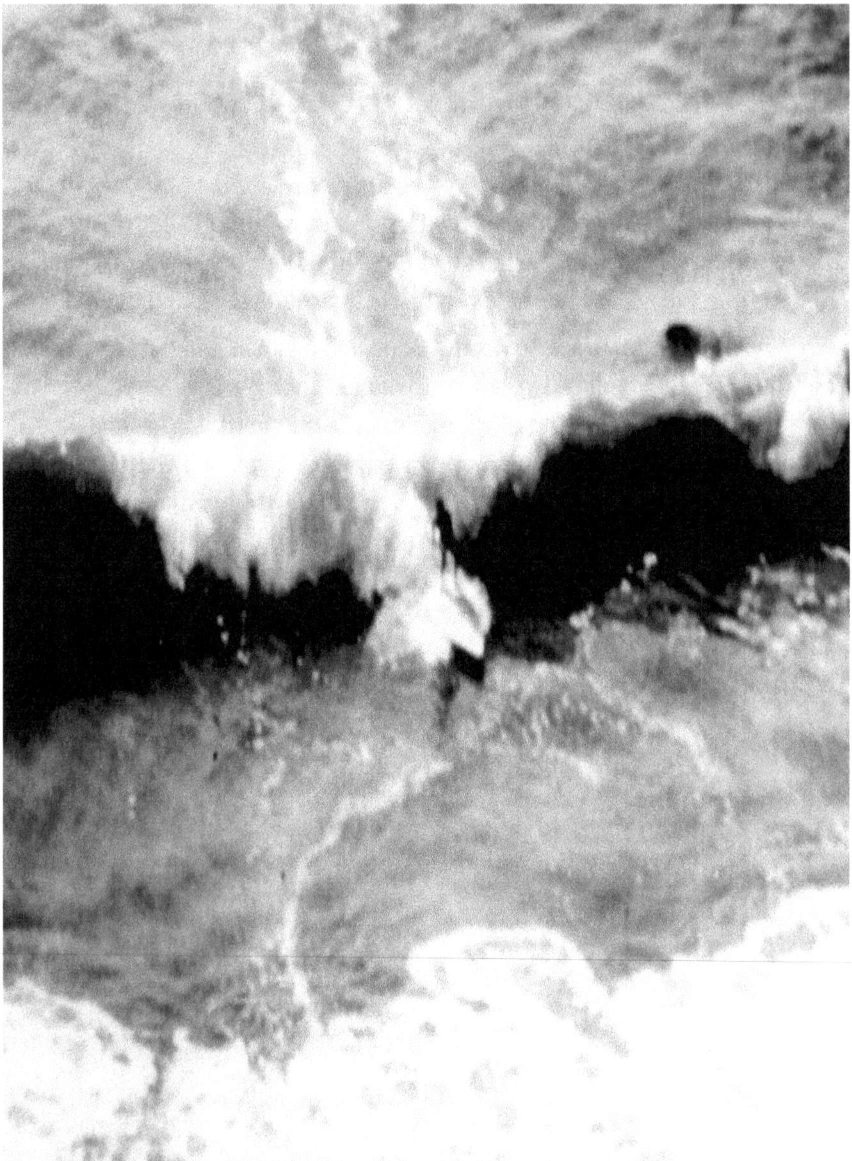

Leon multi banana boat. When you understand in reality the magic of life, you can create magic. When you believe in blind-faith, you cannot create in the dark.

I worked in Haifa, Israel as a lifeguard helper, during my first year.

I always went to work early in the morning because I would rather be on the beach than anywhere else. I took care of everything because the chief lifeguard showed up late and relaxed the rest of the day. One Friday night, I went to a party and got drunk because of an argument with my girlfriend, Susi. The reason for the discussion was that Susi came to Israel to get married and have children. She offered me $500,000 if I married her and gave her a child. She said if I didn't want to stay married, I could leave and keep the half-million dollars for myself.

Disturbingly, ever since the paralyzing injury when I was a small boy, my family has been interested in my money—from the accident insurance, and my lifeguard pay. In the summer, a lifeguard on my beach usually works 14 hours a day, with no days off. As a result, I accumulated double overtime and triple overtime on my paychecks. I was making five times the salary of my father and three times the doctor's salary, who worked for the Israeli government in social health care. All their desire for money kept me away from my family.

When Susi made her generous cash offer, she thought I would jump at it, but I didn't. Because of Suzi's argument about marriage, I drank too much alcohol and couldn't sleep at all. I had a hangover in the morning when I went to the beach to open the lifeguard station. I raised the black flag, which meant no one was allowed in the water. I felt that I could not be alert enough to be a responsible lifeguard. Finally, at nine o'clock, the other three lifeguards showed up. I told them I was very sick and didn't sleep all night and would like to lie down. They all responded, "Leon, you always take care of everything for us. Don't worry—we'll take care of you. Go to sleep." I put a blanket down and passed out on the beach. Soon, my girlfriend Susi came and sat by me.

A few hours later, while I was asleep, three young girls were drowning. The lifeguards at the station called me for help with the bullhorn. I shot up like a bolt of lightning and ran toward the rip current. I picked up one of the girls, moved her out of the current flow, where she could stand safely, and instructed her not to move. I went back and rescued the second girl and handed her over to another lifeguard. I dove under the water to get the third girl. I pulled her to shore, where she started throwing up; at that moment, I knew she was okay and then went back to collect the first girl at the edge of the current. My concept of being a lifeguard is calculating how to save everybody and never give up on anyone. Because every Life is precious and equally important. My friend Joseph from the beach restaurant greeted me with a cold beer when I came back. He said, "Leon, you were incredible! You ran so fast that I couldn't see your legs. This one's for you!" I said, "No, thanks. I don't want any beer right now. I have a headache and I'm not going to drink anymore."

The next day was a cold weekday. Nobody was on the beach, so I sat at the beach restaurant with my girlfriend Susi. We enjoyed ourselves. Joseph, the owner, screamed at me, "Leon, the Mayor of Haifa is here! Get up!" I didn't believe him, but sure enough, the mayor was there, and I said to him, "I'm sorry that I'm not sitting at the station." The mayor said, "I'm not worried. You can be sound asleep and still save everybody. I heard a lot about you. I'm firing the chief lifeguard and hiring you to be in charge." Usually, you must begin as a helper lifeguard, work your way up to the second chief lifeguard position, and then work several years more before becoming a chief lifeguard. It takes approximately twelve years. I became the chief lifeguard of the public Haifa beach within my first year. Love what you do, do more than you think you can do, and by free will, make your Life challenging and exciting.

On duty, one day at the lifeguard station, I noticed a large woman on the beach walking into the foaming surf. Suddenly a strong wave knocked her down and submerged her. She stood up, calling for help. I immediately rushed out to see what was wrong. When I arrived, she screamed, "My baby, My baby! I can't find him!" As I approached, I did not see any baby in the foamy water; the woman was so big, and the child was small. In a panic, the frightened mother said the wave knocked her child out of her arms into the murky water. I quickly searched as hard as I could but could not find the baby. I called everyone in the area to hold hands and feel the dark water with their legs for the little one. About five minutes later, which is a long time to be underwater, someone found the infant.

Of course, the mother wanted her baby right away, but it was critically important to give the child some air immediately. I grabbed the baby like a football and ran into the lifeguard bungalow, where I kept my surfboard. I closed the door and locked it from the inside while I frantically started artificial respiration on the baby. Outside, a crowd was banging on the door, trying to break it down. However, to my relief, the baby finally regained consciousness. As I handed the tiny survivor back to his mother, **I realized, if the baby was not alive, the baby's relatives might have killed me!**

This rescue was reported in the Israeli newspapers and then later in a uniquely American book that celebrated Israel's 50th Anniversary entitled "An Eyewitness Account." (I have included part of the article and the photographs in this book).

To get to my lifeguard job, I had to commute for about an hour on a train that passed by my beach, but there was no train station there, so I had to continue to central Haifa. Then I had to hop for a ride for twenty minutes back to my beach in a diesel smoke-belching bus. Because I hated the smell of diesel so much, I decided to jump off the train as it slowed when passing my beach. Near the coast, the train would slow down from eighty to ten miles per hour for safety reasons so that bathers could cross the tracks to and from the beaches. People would come over early to the coffee shop and watch me jump off the train. It was like being in the circus.

I used to visit my sister in the city of Akko, which was about nineteen kilometers away. I ran that whole distance on the beach since I enjoyed running and hated taking the bus. It took me about two and a half hours each way. I had to go around the harbor to get to the beach. I would be fully awake once I started running for about one kilometer. Then I would fall into a kind of trance until about a kilometer before arriving at my destination when I would snap out of it. It was the same on my way back. This feeling of euphoria is known as "runner's high." Sometimes I encountered a flock of seagulls that would wake me out of my trance. When I tell my brother-in-law that I used to run from my lifeguard station to Akko, he does not believe me. He thinks it was just too far to run.

Being a lifeguard, I learned mouth-to-mouth resuscitation and felt it was wrong because many people drowning often chokes on a floating object caught in their throat. There is no way to see inside the mouth, so you can never be sure. I decided to research anatomy with a doctor friend and found out that air blown into the nose goes directly to the lungs, while air from the mouth goes to the lungs and stomach. If I hold one hand over a victim's mouth and blow air just into the nose directly, then the air would not escape from the mouth and would bypass any choking objects. I could then blow rhythmically into the nose to control the airflow. I knew my mouth was always bigger than anybody's nose, making it a lot easier and more effective to save someone. Once I blow air into the nostrils, I pull the drowning person's knees toward their stomach, collapsing the lungs and pushing air and water out. Then I bring the knees down, blow the air back into the lungs through the nose, and maintain rhythmic breathing. I hope this will be educational so someone out there can be saved.

There was no work in the winter because the beaches were closed, so I went to school, studied massage and physiotherapy, and then got a job as a physical therapist at the Dead Sea Clinic. This knowledge helped me become a better lifeguard because I learned the anatomy of people.

YOYNAY

From time to time, I worked on different beaches: Haifa, Kfar Vitkin, Caesarea Hotel, Netanya and Vinget. Vinget is a very isolated place with a school for athletics' teachers. When I worked there, I met a man by the name of Yoynay. He lived in a cave-like trench he dug into the beach covered with metal signs and sand over it. He lived in that cave with twenty dogs. One day, Yoynay told me the story of his life.

He was a Danish Christian who fell in love with and married a young Hebrew woman. After the Holocaust, she wanted to live in Israel. However, once they arrived, they could not adjust to living with the Hebrews, so they moved to reside with the occupying English Army that controlled the country. To survive, his wife had to sell her body for money to the British forces' officers.

One day, a Muslim Sheikh wanted to pay for her services. She said she would not have sex with a Muslim because they do not respect and treat women fairly. Later, as she traveled with her husband in a carriage, a gang of Muslims stopped them, pulled her out and slit her throat.

Yoynay was very disgusted with all the people at that moment. He decided to live in isolation with his dogs. He lived off the fish he netted with the help of his well-trained animals. On one side of the bay, he would enter the water and walk toward the other side where his crafty canines were waiting, while simultaneously extending a large net with floats on the top and weights on the bottom to corral the fish. Then he would signal his dogs to chase more fish into the net, which proved very useful. He told me he would rather live alone with his dogs than live with people.

Leon in a, resuming action

ENGLAND

Before going to America, I was invited to stay with my friend Donald in England for six months. I met him during my acrobat performance days. Donald Bowden was a member of England's royal class. He took me every day to visit some royalty where we had tea and crumpets. Donald always introduced me to their highnesses as the "Prince of Tripoli." I listened to these royalists talk about the peasants of England and how inferior they were to the upper class. All England's royalty spoke Franch.

One day a truck backed into Donald's brand-new Rolls Royce and scratched the top of the hood. I said, "Why don't you ask the man for his insurance?" Donald said, "It's okay; he's just a peasant. **The year 1964.**

On another day, he showed me Great Britain's Queen Elizabeth II on television. Donald asked me, "Isn't she beautiful?" I said, "I was a lifeguard on the Mediterranean coast and have seen and been with lots of beautiful women on the beach. She is not like any of those beautiful women,

One time, we were invited to visit a Baron, who asked me, "Do you like my driveway with all the colorful pebbles on it?" I said I did. He said, "When the loose pebbles were initially laid down and when my children drove their Jaguars too fast, they spread them all over the place, so I hired peasants to come to glue the pebbles down. Don't you think that's a great idea?" Sarcastically, I replied, "Aren't you all so lucky to have peasants do beautiful things for you! We don't have peasants where I came from." Everybody laughed.

Another time, we were invited to the queen's palace for dinner. During a conversation, someone said, "Do you know this table is worth a million pounds?" I asked why and he said, "Because King Henry VIII wanted this table so much that he went to war, resulting in thousands of deaths just to obtain it." I added, "As a lifeguard who saves lives, when I hear how many people lost theirs, I wouldn't pay a penny for this table because it is a bad omen, and it is not humane." He got furiously upset. When the party was over, everybody went outside to the lawn, where there was a bit of white wall for guests to sign their names. I signed my name in Hebrew. In the end, I respect the queen because she is a good representative of her people.

I had an eighty-year-old letter that referenced a 280-year-old letter, which stated that ancestors on my father's side helped George Washington win the American Revolution. When I showed this eighty-year-old letter to the American consulate in Israel, the man who reviewed it was amazed. He said, "This is beautiful history." As a result, I obtained a visa and a green card from the consulate and that is how I was able to go to America. (A photograph and transcript of the letter appear in a chapter below about my father's ancestors.)

AMERICA

After six years as a lifeguard, I decided to leave Israel for good and emigrated to America. I left my accident insurance money, my lifeguard money, and my friends to go to the land of Mason's wisdom. My father told me he would hold these funds until I needed them. Unfortunately, I never saw any of that money again. When I arrived in America, I had only one hundred dollars in my pocket.

It was challenging for me, because I could hardly speak English and didn't join a group or belong to a temple or a church because I always do what is right and will always be a free self-thinker Hebrew in a search of a universal truth for a utopian world. I could not get a good job; I didn't assimilate quickly. It was hard for me, especially after my easy life on the beaches, because if you understand right from wrong, you will do what is right and what is right is to stay, so I took it as a challenge. And after a while, I made great new friends. We would talk about anything. People think because they are from different origins and backgrounds, they are not able to communicate. That is not true. All you must do is stop being a believer, and instead search for "universal-truth" for the time. By debating in God in reality of science, for more accurate understanding of the spectrum of the magic of life, you will have an open, truthful mind to know right from wrong to be a humane-being, by the people, for the people and from the people! Sticks and stones will hurt your bones, but words will never hurt you. In a conversation searching for universal-truth as a humane-being, you will always have harmony.

Even though it was hard starting a new life in the USA, I never wanted to move back to Israel. I permanently moved forward. Life is a journey of challenges. In reality, I could not get a job as a lifeguard because I was not an American citizen, so I had to do something else. After arriving in New York City, I began teaching children acrobatics for three hours a day at the YMCA. I showed them the moves and before you knew it, they started bringing their mothers to watch, and it became a circus show!

At first, I had no money and had to run after the bus because I did not know how to get to the YMCA. My friend George asked why I ran. I responded, "I like running! I don't like the diesel smell on the bus. Plus, I save a dollar each way." George said, "If that's the case, why don't you run after a taxicab to get to and from the "Y"? Then you'll save ten dollars each way!"

One day I was hungry and broke. I entered an Italian restaurant and told them I would work for food. The manager said, "Okay, do some work in the kitchen." There was a big pile of dishes in the sink, so I started washing them. When I was done, the manager walked me into the dining room and placed a big plate full of food on the table: spaghetti with sausage, salad, and dessert. It was delicious! When I finished, he gave me twenty dollars. I did not think I deserved twenty dollars. That was a lot of money in those days, equal to about $100 today. This was as much as I had done per 1 day at my YMCA job. I protested, "This is too much." He said, "People come to eat here. Those who cheat or don't have money run away without paying. You are the only one who offered to work, and you did an excellent job. The twenty dollars is for making me understand that there is hope for people to become humane-beings in the future."

I did not get along well with American women. In some places in the world, men mistreat women, but in America, women mistreat men. For example, before a man has sex in America, the woman tells him that it is his lucky day. I never heard that in Europe. However, I do not want to mistreat anybody, and I do not wish to be mistreated; so, I wrote to three of my girlfriends in France and asked them to visit me in America. I did not expect all three of them to come; however, they arrived within three weeks, and stayed with me. We decided it would be nice to travel around this great country. At the time, I was in business with a cabinetmaker named Henry. I sold my half of the company for travel money.

I invited two more friends from Belgium to come with us on our trip. We bought a huge car, a green Oldsmobile Cutlass 88 convertible. We loaded the tent and the rest of the camping equipment in the trunk of the car and hit the road. Gas at the time was 25 cents a gallon! We started in New York and ended in Florida, staying at campgrounds along the way.

While camping in Key West, Florida, we went fishing every day. Then, in the evening, I barbecued fish, made conch chowder and bouillabaisse, and invited everybody in the camp to eat with us. They were nice and friendly and brought us wine and other alcohol. One person I met was a firefighter on vacation. He said, “I was supposed to stay here only for two weeks, but I’ve been here for two months! So, I’m going to stay until you guys leave because I’m having such a wonderful time!”

In his mobile home, I saw a plaque on the wall that read: “The Bible is the Word of God!” I said, “I can prove to you that the Bible is not the Word of God. I know the Hebrew people who wrote the Bible. Since the Bible is made of words, they can be misinterpreted and people make mistakes, but God does not.” After I said this, the fireman got mad. He grabbed me, kicked me out of his mobile home, closed the door, and drove away. I had made one comment that was critical of his beliefs, and he became a monster. He could not have a conversation because he was a believer in blind-faith and refused to accept evidence against his blind-faith, called cognitive dissonance. He had not learned the scientific art of debate-conversation in search of universal-truth to understand reality by free-self thinking and debating for a utopian life and live life with a smile.

One day, a young man asked me to teach him how to spearfish. So, we went out on his motorboat with my three French girlfriends. We anchored near a rocky area. The women swam near the boat while the young man and I dove toward a cave where big bass fish were swimming inside. I struck the outside of the cave and shook the water with my hands to attract the fish, but it would only come out halfway and then go back inside. It frustrated us both because I could not spear him. This was not normal, and I had a feeling that something was wrong.

Perhaps the fish knew something, I did not. Then I raised my head and saw a giant shark coming toward me like a torpedo; it was about twenty feet long. Even though the water was crystal clear, the killer shark was so giant I could not even see its tail moving. The shark's head was as large as an old Volkswagen. I was terrified; it was a Great White! My swimming partner had not seen it yet. I raised to make myself look more significant and pointed my small speargun at his eye. Mentally, I was telling him that I would not go out quickly.

The shark approached ten feet from us and stopped. He stared me down—and finally turned is head to swim away. I signaled the young man to swim to the surface. Then I called out to my French female friends, "Man-eating shark! Get on the boat now!" On the way back, we saw many other sharks, including hammerheads. Once we experienced this shark migration, I decided to leave Key West. I was not going back in that water! Over the years, sharks have attacked me four times. I learned that sharks—like most animals and children—try to get away with as much as they can, but they usually don't take chances if they think they may get hurt.

The next day, we left for New Orleans and had a lot of fun. I got along great with the people there since they all spoke French. The music was hot and so were the women. When walking down Bourbon Street at night, I learned that if you threw a necklace of colorful glass or plastic beads to a willing woman, she would lift her blouse and expose her lovely breasts. America, what a country! A week later, we left and headed to Texas.

Leon and his three French girlfriends, exploring the United States for eighteen months in a tent.

We could not find a campground there, so we set up our tent on top of a grassy knoll in a public park. The police came and said, “Hey, you guys, you can’t camp here!” So, I told the three girls to open the tent's zipper and stick their naked legs out, can-can style. The policemen’s eyes bulged out when they saw the lady's lovely legs and said: “Okay, you guys can stay, but you have to leave in the morning.”

On the way to Brownsville, Texas, another French friend named Louie Nitka joined us. He helped with the driving and pulled up behind a police car, which was moving very slowly. Surprisingly, Louie bumped the police car from behind. The cop stopped and got out of the car with his gun drawn. He asked, “What do you think you’re doing?” Louie explained, “In France if a car goes very slowly, you’re supposed to bump it so it will move out of the way." The policeman said, "Not in this country, you don't! Because this country is a country of law!" The police let us go with just a warning.

When we arrived in Brownsville, Texas, we stayed at a motel. Together we went to the Mexican consulate to get permission to cross the border. They gave everyone permission except for me because I had an Israeli passport. The official said, "If you want to enter Mexico, you'll have to give me eight hundred dollars. If you don't make any trouble, I will give you the money back, **because the Israeli government, kidnapped the criminal Eichmann whiz out the permission of the country."** I said, "No way am I giving you eight hundred dollars. You don't look like you can be trusted! If I wanted to, I'd be able to get into Mexico and you would never know it." So, my friends spent some time in Mexico without me. Meanwhile, I got hired to work on a shrimp boat in Port Isabel, Texas. All the shrimpers there speak French, and they will sink your ship if you don't speak French.

One Saturday morning, I entered my motel's lobby and saw that it was full of people. I asked the manager what was going on. He said, "Mexican girls come over the border on the weekends to make money." As I was walking around the lobby, a Mexican girl in her twenties called me over and said, "For you, exceptional, two dollars." I said, "No. I never pay for women." I walked by a little later and the same girl said, "For you, very special, for free." I said, "No, I'm okay." She followed me around and said, "For you, very special, I give you two dollars." I laughed and we became friends, and she lived with me for a while.

I met a Mexican-Indian friend and told him about my plan to go to Mexico. "I'll dress up like a Mexican, grow a mustache, and act as a mute around the local police—you'll talk for me. Mexican mutes don't need papers." He agreed and we traveled together. He took me to his family's place, where there was a whole tribe of Indians. I found out they did not let their girls marry unless the men were from their tribe.

I felt a pang of disappointment that I was only a Mexican disguised because the young women were beautiful.

While walking on the street one day, an older woman carrying a big bag over her head tripped and fell. I ran over to help her get up. A priest came over and thanked me. He had a French accent, so we started speaking in French. He wanted to show off the power he had over the people since everybody respected him. I knew that Mexicans were very religious, fearing their religion and their parish priests. (God-players.) While the priest and I were walking together, we saw an older man selling one watermelon and many tomatoes. The priest picked up the watermelon but did not pay for it. I said, "How can you take fruit from this poor old man without paying?" He said, "They love to give to the church." I did not think that was right, so I said, "I don't want to associate myself with you anymore."

Some people nearby must have heard what I said to the priest and didn't like it. When I arrived at my hotel, many young toughs came up behind me and tried to beat me. I could not understand it. I defended myself, but there were too many of them. I was fortunate that my Indian friends arrived just in time to save my life. After this event, I decided to leave Mexico and head to California.

I met my French lady friends from the trip at a bus station in Phoenix, Arizona. We drove west together to the Leo Carrillo Campground in Malibu, California, and set up camp. The next day was Saturday. I got up, went for a walk along the beach on the south side, and made my way into a restricted area named Nicholas Beach. I saw a group of people pushing a sailboat toward the ocean. Looking at the boat, I realized the sail was too big for the size of the boat. I said, "You know you cannot sail this boat safely. It's going to tip over." They all laughed and continued to go sailing.

As soon as it passed some waves, the boat tipped over. Because I was formerly a lifeguard, I jumped in, swam toward them, and helped turn the boat upright. I said, "The only way to sail this boat is by going against the wind. Tack back and forth—just go in and out. If you go with the wind, you won't be able to come back. So, they continued to sail on successfully.

The owner of the boat, Bob Green, whose house was on the beach, saw me flip up the sailboat and rescue his friends. Among others on the beach who saw what I did were the actor/comedian Paul Lynde and Bob Hope's son. Bob Green came up to me and asked, "Where is your mother?" "In Israel," I replied. He responded, "Let's call her and tell her you're okay." He was kind and must have thought I was a teenager, but I told him I was the man who just saved his boat and everybody on it. We laughed together. He invited me into his home, where snacks and drinks were served, and asked me where I was staying. I told him I was staying in a tent at the Leo Carrillo

Campground. He said, "Why don't you bring your tent over here and stay with us?" I gratefully accepted his offer.

One of his friends asked if I knew how to play backgammon. I said I did, and he said, "Let's play for money. How much you got?" I said, "Three hundred dollars." He said, "Let's play for three hundred dollars, then." I said, "No, it's the only money I have." He said, "Let's play for a hundred and fifty.' I said, "Okay." I do not like to gamble, but I knew I was an excellent backgammon player. It is like taking candy from a baby and he was wealthy. We played on and on and I kept winning. Finally, I was ahead by five hundred dollars, and he was getting upset. I said, "I don't want to play anymore if you're going to get angry." He said, "Let's play double or nothing. If I win, I pay you nothing; if you win, you get a thousand bucks." I said, "Last game." He said, "Okay." So, we played, and I won a thousand dollars, which was a lot of money I could use at that time!

Later that day, the woman who went out on the sailboat with the oversized sail came limping towards us and said, "The boat is stuck in the rocks two miles south down the coast! We can't bring it back!" I reminded her that I said not to go downwind and sail against the wind, tacking back and forth. She asked me, "Leon, are you going to help us?" I said, "Yes." So, several of us headed out to Bob's sailboat for the rescue. When we got there, the sailboat was pinned in waist-deep water and banging against the rocks. After reviewing the situation, I said, "We can't carry this boat back. It's too heavy, too far away. The only way is to sail it back." They suggested I take someone with me. I said, "No. It's too much responsibility if something goes wrong. I have to do it by myself."

I sailed directly about two miles into the ocean, as far north as possible. The people on the beach couldn't see me anymore. Bob Green called his neighbors and the Coast Guard for help because he thought I would never return. I had to go far enough away from the shore to make only one tack directly toward the house on the beach. Otherwise, every time I tacked; the boat would go downwind away from the home because the sail was too big. Finally, I turned the boat around and sailed directly toward the house. Approaching the beach, I caught a wave and surf-sailed the boat back to shore. It was sunset when I arrived. There must have been two hundred people on the beach, and they were all applauding me. All the neighbors were there to see the event. Because of my heroics, I instantly made friends with Bob and all the people on that secluded beach. Among those who witnessed the rescue were such luminaries as actors Steve McQueen and Vincent Price, Secretary of State Roy Ash, Jack Brady (who later became my very best friend), and two young men by the names of Gerard Alcon (a French Basque) and Robert Bardey (a French commentator). I first met Gerard and Robert, who were newscasters in America for the Voice of Europe. As mentioned in a previous chapter, we became close friends since we could converse together in French.

CHAPTER 45

THE SECOND GUN

On one of the darkest nights of my life, the three of us were at the Ambassador Hotel on June 5, 1968, covering a vital presidential campaign speech by Robert Kennedy, when the popular, young Senator from Massachusetts was assassinated. After his speech, Kennedy and his entourage exited the stage. We were still filming the Senator's departure through the hotel's pantry when multiple gunshots rang out. The fantastic thing is that some of our footage and subsequent interviews reveal what happened that night, including what the medical pathologist who performed the autopsy had to say. We produced a feature-length documentary together and called it *The Second Gun*, which was reviewed and written about in the *New York Times*, the *Los Angeles Times* and many other newspapers and news outlets worldwide. We distributed the film in theaters domestically to show that there was a conspiracy. Today you can watch it on YouTube.

About a year later, while shooting a new film on location in Utah, my good friend and associate Gerard Alcon, co-producer of *The Second Gun*, became mysteriously ill and died within a week. My theory is that he was assassinated by poisoning his food in the restaurant. I took care of him for the week he was suffering before he died. His doctors were stumped. They could not identify the particular poison that killed him and without an antidote, they could not save his life. Although I could not prove it, I had a powerful feeling that he had been killed; he ruffled so many political feathers with his controversial conspiracy film. But, as in the past, hierarchy does not want the truth. Gerard was an American hero. He insisted that Robert Barley's, and my name as co-producers would not appear in the film credits because he did not want us to get hurt. Gerard said, "I am glad you are alive, so, keep on searching for truth for a better world. A few years earlier, Gerard was also one of the first reporters to uncover J. Edgar Hoover's dirty little secrets, turning Gerard into a target. I respected him and missed him.

I was invited to stay with Bob Green, who wanted to build a shopping center in Malibu. He asked if I would like to be involved, and I jumped at the idea. While waiting for the architectural plans, he hired me as a lifeguard on his secluded, private beach to make sure no one drowned there. I served on that beach for about a year and kept everyone safe.

One day, Jean Buchanan, a friend of Bob's, came to visit the house. When she got out of the car, her body was hunched over and folded like a pocketknife. She told me that she was going to have back surgery soon. Since I was a trained physical therapist, I asked her, "Why don't you give me a chance to fix your back? Give me two weeks and if I don't fix it, you don't have to pay me anything." Back surgery was hazardous at that time and not consistently successful—that is why I wanted to help her.

I started her therapy with stretches, gravity exercises, and massages. She felt great after a week. She said her doctor wanted me to work for him and would pay me handsomely. I told her I was too sensitive around sick people to work full-time as a therapist but would always be willing to help someone in need. Also, I was about to open a big shopping center with Bob Green. We planned to build ten barns with distinctive designs from all over the world on seven and a half acres of land adjacent to the Pacific Coast Highway near Malibu. Each barn would feature different products to sell.

To enhance the appearance, we created a small pond from a flowing river through the grounds. In addition, we designed the area so customers could park their cars outside and walk joyfully into the shopping complex.

I lived on that property in a mobile home, which was right next to the Pacific Coast Highway. I worked every day with a tractor to clear the land. One day, I was cutting bushes with a machete when suddenly a six-foot rattlesnake jumped up and struck my face. I jumped back quickly and cut its head off with the machete in midair. That was one very close call.

As the project moved ahead, Bob Green and Jean Buchanan kept spending all their money on parties. Finally, we ran out of cash for the project and eventually, everybody started suing each other. I had to get out of there.

CHAPTER 46

FHA FOSTER HOME PROGRAM

At one time in my life, I felt that I should take care of neglected children, although I didn't want any offspring of my own. It was not essential to have someone call me "Daddy." I found work in a program with the Federal Housing Administration in Sun Valley, California, tasked to take care of 350 foster home children. President Lyndon Johnson created 500 of these projects, because after the Vietnam war, many American fathers were killed, there is no wisdom in war!

Once I moved in, I realized children behaved according to Nature's law, i.e., like wild animals. The children would fight among themselves continuously in gangs. They were so violent they would even kill one another! I confronted them one day and asked, "Why don't you be nice?" They replied, "Nobody is nice to us." Some of the children added, "Our parents gave us away to foster homes, but after a while, the foster homes didn't even want us anymore. So why should we be nice?" I said, "When you are nice, you have nice friends. It's worth it." The kids said, "We don't want to listen to grownups anymore. Some people are still childish, like these foster home kids, and they think they know everything. No matter how ignorant or fanatical religious people lack understanding, problems arise because they refuse to have scientific conversations in wisdom to create an accurate understanding of reality. Like in the wisdom story of the Bible, I performed magic tricks to get their attention so they would listen and get the children involved in scientific style debate conversations to understand life.

The next day I installed a rope on the roof near two big spotlights shining on the parking lot below. After dark, I climbed on the top and used binoculars to spy on the kids. I could see them, but they couldn't see me behind the spotlights. While checking on the kids, **just like Moses, climbing up the mountain checking on his people,** I saw many things they were doing at night. On one occasion, I saw a youngster named Johnny break into a Cadillac and destroy the dashboard to steal the radio. The next day, I said to him, "Hey Johnny, last night I dreamed about you. It was not a nice dream. You probably don't want to hear it." Johnny said, "No, no, go ahead! Tell me about the dream!" I proceeded to tell him what I saw from the roof but emphasized that it was just a dream. Johnny asked me, "What if what you dreamed happened?" I said, "If this is what happened, God, the Cosmos is upset with what you have done." Johnny started laughing. He asked, "God? What is that?" I said, "God, the Energy of the Universe, is worried about the way you are living, and he wants me to help you; that is why I had this dream." Johnny said, "What? Tell me more." At that moment, I hooked him and the other children with my verbal calisthenics, encouraging them to say what they feel but listen to others too in a civil conversation. That's how you learn from other people. They were happy when I told them, "Today, I will fix the maintenance room and install a pool table. You and your friends can help and debate for a more accurate understanding of God's universal truth, by accepting only the alliance between God and you, because once you understand in scientific reality, there is a Creator, no one can demoralize you or make you immoral.

The kids enjoyed coming to the new game room. They began trusting me and wanted to know more about what I had to say. I explained by debating. I am here to help you free self-think, so you can understand more accurately right from wrong in the magic of life reality, and **I will not preach**." I debated with them the book of the Masons. I explained, "The Energy of the Universe is the biggest builder. It built the stars, the sun, the moon, and the Earth to live on a planet with a "Goldilocks" atmosphere, not too hot and not too cold.

All people are children of God. Those who work the hardest to create a better world for all life are godlier than those who do not. The Energy of the Universe gave you life, so use it. Being alive is a better gift than anything anybody ever gave you or ever will give you. Do not blame your parents or anybody else for your disappointments. When you work hard, you make your body and mind more robust and healthier, and you learn new things. When you use your mind and free self-think, you become more realistic and creative. When you work hard as you get older, you become wiser, healthier, and happier and that prevents Alzheimer's sickness. So don't waste your time complaining. Challenge your life to overcome life's obstacles. You have nothing to lose. Everybody eventually dies anyway.

If you want to understand energy of the universe, the creator, by free self-thinking and in a scientific debate, find the wisdom to understand in your mind in scientific reality, to do what's right, and enjoy the fleeting time you have on Earth because life is precious and shorter than you think. When you overcome the most significant challenges, it becomes easier to handle new ones. Everyone can learn and evolve for the better, make the right choices, and improve the world for all lives. I know I tricked the kids, but I did it because I wanted to help them learn from a scientific debate in conversations of reality to ask questions of themselves and each other. Within three months after I started helping the foster children move forward in life, the youngsters got jobs, went to school, and were not in gangs.

One day, eight police cars drove up to the project and one officer asked, "Who is Leon? We want to talk to him." I answered, "That's me; what is this all about?" I was apprehensive that with all these policemen here, something terrible must have happened. So, I invited them in, and they said, "Three months ago, these kids used to be in trouble all the time, but now they're not. They've changed. They're such nice kids. How did you do it?" I told them, "It was a life magic reality!" We all had a good laugh.

A year later, an inspector from Washington, D.C., came to visit and said to me, "You know, out of the five hundred foster homes projects we examined—this is the only one that worked successfully. I want to promote you in charge of overseeing all our projects." "Thanks, but no thanks," I replied. I am a young man. I'd rather go surfing and have fun on the beach because I love to enjoy life. I only wanted to help as a contribution to the kids To prove that free self-thinking in a scientific method of debating creates a more accurate understanding of God's reality, the reality is not an illusion And accurate scientific debate in the search of universal truth always works.

It was not their fault they were foster children; it was society's fault because this society is based on competing for the hierarchy that created freedom of illusion of knowledge (religion) in blind-faith, to follow a hierarchy with knowledge but with no understanding and competing power over others, by giving people, shelter, food, protection, to follow as **sheep in the herd of grazers, that becomes a common belief in blind-faith, to protect hierarchy privileges.** It is time to change for the better, because people are not sheep, and can become brilliant humane-beings, by free self-thinking in search of universal truth, by debating in God's reality of science, for a democratic utopian life.

Thirty years later, one of my former foster children from the project called and asked if I was Leon from the Sun Valley Project. I said yes. He said, "You don't remember me, but I used to be one of the kids there. My name is Johnny, and I want to visit you." I gave him directions and after arriving at my house, he presented me with a gift: two beautiful replica swords of the Knights Templar. Of creativity, and wisdom. He said, "You taught me the wisdom of the Masons, which helped me very much and now I am a successful real estate broker. To show you my gratitude, I needed to find you and give you this gift. If you ever need any money or help of any kind, call me." I said, "The swords you gave me make me very proud of you. But the best gift of all is that you turned out so well. Keep on free self-thinking in wisdom to create a better world, for all lives."

The next photo is of copies of the original swords of the Templars. The King of France captured the two Night Templar representatives of the Masons on behalf of the Roman Catholic Pope, who judged them for being liars, therefore they are heretics, and he condemned them to be burned alive at the stake. The King and the Pope are examples of all-powerful hierarchies that commit horrendous crimes. Look at the Spanish Inquisition, Nazi war crimes against life, China's re-education camps, Russian civil repression, and their prison gulags in Siberia. The Mason movement implies **you are what you do, not what you believe you are!** Because all the land was conquered by extreme violence, however, today that is acceptable as a universal-truth. **So, please don't tell me that religion makes you moral!**

If you are creating a better world, you deserve land and the resources you need, to make more of a better world for all lives! For example**: Jack Brady had to move, to create a public beach. Harold Cremer had to move to build a highway. Hate and wars are not the answer.** It is essential to create fairytales to help children follow the right path towards becoming humane grownups. But it is counterproductive to develop illusory religious and political propaganda fairytales for adults because adults need *accurate factual realistic truth* for the scientific debate conversations necessary to develop new knowledge, **because people create all great civilizations, and greed from hierarchies destroys all civilizations!**

Some people tell me that wars are necessary to lower overpopulation. My **response is, "**Why don't you kill 2 birds with 1 stone? You can start by killing yourself. This way, (1) you will lower overpopulation, (2) you will get rid of inhumane people on this beautiful planet!"

This beautiful gift is from kids of the FHA foster home program.

Replica Swords of the Knights Templar

CHAPTER 47

DESILU

I was hired by Desilu Studios, where I oversaw the **forty-acre backlot.** The studio produced television shows such as *I Love Lucy*, *Hogan's Heroes*, *The Andy Griffith Show*, among many others and numerous TV commercials. It also housed standing sets from films like *Gone with the Wind* and the original *King Kong*. I met many actors and animal performers there (Although some of the actors on set sometimes acted like animals). Later the studios began losing money and were eventually sold, torn down, and replaced by commercial buildings.

One Monday morning, while the studios were still operating, I arrived at the front lot where a vast commotion took place. There were police officers and detectives around with lots of onlookers gazing through the chain-link fence. I asked one of the detectives, "What happened?" He informed me that some vandals broke into the studio during the weekend and caused half a million dollars' worth of damage. He said, "We are taking fingerprints and looking for clues to catch the culprits." Then I said, "Watch me catch the culprit in five minutes." He laughed at me, but, by looking at the damage, I realized this vandalism was children's work. .

I strolled over to the chain link fence and observed the crowd. I noticed a child who was about ten years old, moving around nervously. He looked guilty to me because his fingers holding the fence were trembling. I walked up next to him, grabbed his hands on the fence, and proclaimed, "Here is the kid who did it!" He screamed back, "It wasn't me; it was my brother who did it!" As the detective approached, I said, "See, I told you I could solve the case in five minutes." He was surprised and laughed again. Sometimes, you must think for yourself, realistically, in wisdom, and creativity, and not have to go through tedious knowledgeable procedures.

Often youngsters would come to the studio and ask me if they could visit. I would let them in and show them around. But I would also catch kids who would break in and damage property. So, I would take them to the Los

Angeles River, on the west side of the studio, and tell them, "I will let you go, but you have to run fast. If I catch you, you'll be sent to the police department." I also told them that they have to ask if they want to visit, and I would try to accommodate them. Once the children were on the sidewalk, they started running as fast as they could. I was running fast, too, but I made sure to keep a foot or two behind them for at least a mile. Those out-of-breath kids never broke in again. While working at the studio, I had an idea and told the studio manager, John Pepin, that many film fans and collectors might be interested in the historical objects, props, and costumes we had in storage. We had memorabilia from the original *Tarzan*, and the original *King Kong*, *Gone with the Wind*, and other classic films. I suggested that we could charge money to let people come in and give them a tour. Our rival, Universal Studios, started opening its gates to the public about a year later and charging money for a tour of Universal Studios.

My girlfriend Nicole at Desilu Studios

CHAPTER 48

LEON'S MANY FRIENDSHIPS

When I finished working for Desilu Studios, my friend Jack Brady got me a new job at Our Lady of Malibu Catholic Church.

THE CATHOLIC CHURCH

I lived in the back of the church as a maintenance man and a janitor. I cleaned and waxed the classroom floors once a week. I told my supplier to get me a cleaner that doesn't attack the wax when mixed. Later I put them in a bottle to spray the floor. Then I was able to buff the floor and clean and polish at the same time. The supplier observed what I did; then, later, using my formula, he started making and selling the product commercially and called it "Spic and Span." He used my wisdom to make the world a cleaner and shinier place faster.

Monsignor Sheridan, the man in charge of the Our Lady of Malibu diocese, came up to me one day and said, "I will be attending a television conference of leaders from all religions. The attendees are supposed to tell what contributions they have made to society. Can you help me figure out what I should say?" I said, "I'll teach you a Hebrew prayer: **"Thank God that we have arrived at this moment where we can communicate through debate conversations."** I taught him how to pronounce it correctly in Hebrew. We practiced for a few days while he learned the prayer in perfect Hebrew.

I asked him to try to be the last one to appear on the show. I watched on television as the religious leaders got up one by one and talked about how great they were by their opinion propaganda to increase their importance and protect their privileges. The last speaker was Monsignor. He stood up and said the prayer in perfect Hebrew. Everybody in the audience and the panel gave him a standing ovation because most understood Hebrew and its message. Monsignor John Sheridan was thrilled.

In Loving Memory
Monsignor John V. Sheridan
1915 - 2010

I spent a lot of time with Monsignor and Jack Brady. We had many debate conversations about the reality of life. Finally, Monsignor encouraged me to write this book. He took a framed picture off the wall, gave it to me, and said, "After you finish writing your book, you are going to need a friend to betray you for money from the hierarchy leader. That will make people furious, and they will want to know what you have to say in your book."

One night I asked him, "Why are you a Monsignor? I know you are not strictly religious." He said, "I'm Irish. The nicest thing a Catholic boy could do for his mother is to become a priest. So, my brothers and I are all priests. Now I have an open mind, and I realize I'm helping a lot of people; that's why I'm still a priest."

During the big fire in Malibu, a giant marble statue of Jesus on a hill toppled over. Since the congregation had planted lots of flowers and shrubbery around the statue, they refused to use a crane to lift the statue because that massive piece of machinery would destroy the landscaping. So, Monsignor asked if I could find a way to return the statue of Jesus to its place. I replied, "I'll see what I can do."

The marble statue weighed a ton! It was also very slippery, so I asked the school kids to help me gather up all the volleyball nets they had. Then I asked them to dig under the statue and wrap the straps around the marble of Jesus very tightly. After mass, I requested Monsignor to ask some male church members to come and help. Monsignor sent approximately twenty men up the hill. They all grabbed the nets, and we lifted the statue onto its stand. As we came down the hill, Monsignor said, "You see, no one else could raise Jesus. It had to take a Hebrew to raise Our Lord and Savior.

On a hot afternoon, I worked on the sprinkler system in the back of the church when Monsignor showed up with a federal Senator. He said, "Saint Leon, I want you to meet Senator John." The Senator asked Monsignor, "Why do you call him Saint Leon and not me, Saint John?" Just as he responded, a black bird crashed on the back of the church window and fell unconscious. The bird was probably dead. Monsignor said, "Watch Leon the Saint." I felt that I had to do something special since Monsignor expected a miracle from me. So, I carried the bird to the church's back room, where there was a vodka bottle in a drawer for the priests. I poured a small amount into a small glass and stuck the nose of the bird in the vodka. Sure enough, the bird woke up. I took it outside, released it and it flew away. Monsignor said, "You see, that's why Leon is a Saint, and you are just a Senator!"

The biggest thing that happened to Jesus was that he was betrayed for Roman money, which upset his free self-thinking friends terribly. Some people treated Roman money as God. This made free self-thinkers mad because they understood that God is not money. They wanted to know what Jesus was saying about his friend Judas, one of his closest disciples, who betrayed him for a Roman bribe. Monsignor Sheridan explained that if Jesus had not been betrayed, he would not be the famous religious icon worshiped today as the son of God. Without such a betrayal, Jesus might have been forgotten, just like the thousands of other humane-beings who were crucified by cruel hierarchy.

Jesus grew up as a Hebrew free self-thinker in wisdom. His upbringing was in the Universal Truth of Moses, not religion or beliefs. He fought the greedy Babylonian Rabbis, the religious hierarchy of his time, by kicking the money bucket! It caused him pain to see an evil hierarchy enslaving, torturing, and killing humane-beings for corrupt cruel greedy leaders, for money, power, and comfort. It reduced people into frightened blind-faith believers and followers of authority.

Jesus did not want to live in a world without love of life. Love is humanist God's universal truth. That was Jesus's plan to make the world a better place. Therefore, he lived amongst as many other free self-thinkers as possible, but they are often not remembered. As we know, Jesus was betrayed for Roman money by a friend! It is happening now, with leaders all over the world who you think are your friends! It is bribery and it is the worst crime because it makes all crimes committed by the powerful forgivable. It is corruption and it creates a jungle out there (Nature's Law). Free self-thinkers, moral humane-beings, have been fighting injustice forever. Corruption and deception are not ethical or humane; they distort your healthy mind, country, and the world because God is not money! God is the universal energy, the creator. ., Free self-thinkers are in the search of a universal-truth and by understanding of the reality of science!

All these great humane-beings from the past are not make-believe fairytales. What matters is the meaning of their message; wisdom changes people into moral humane-beings. They became wise because they understood God's magic of life, from the importance of an accurate understanding of reality. Thanks to their challenging work, we have a better world today. They did it without bribes, believing in religion, or believing in tradition, or believing that there is nothing out there. They were not hierarchy, but they were all free self-thinking representatives of life! They often had to fight against the corrupt leaders of their time. Many would die for their freedom of the mind in scientific search of universal truth to understand that the Creator, the Energy of the universe is real.

I have confronted many different religious people who are certain about what they don't know in their lives. To 'believe' in something means you 'don't know something.' You may believe in Heaven, but you don't know if it exists or not, and if it does, where is it? In science and mathematics, you see the sky is blue, the sun is hot, ice is cold and two plus two equals four. These are not beliefs. These are provable and repeatable facts. But believers act as though they are sure of what they don't know and often try to impose their beliefs and opinions on others. They fight wars to kill or be killed for beliefs, they all are stupidly childish, from blind-faith.

Beliefs are not facts, science, or universal truth. So, all I have to do is exaggerate the silliness of these artificial religious rules and beliefs and they become funny. But be careful what you say to some people. You must first make sure they are not aggressive in blind-faith, or you might get hurt.

Religious leaders, politicians, and dictators have no right to tell us what to think or behave. We must all think for ourselves about what is right or wrong and what path to take in life, not what some "leader" or institution tells us. Otherwise, we shall continue to endure hatred, jealousy, and conflict worldwide based on religious precepts, racial prejudice, or political hatred by opinionated propaganda. To perpetrate such atrocities is not wisdom. It is not correct; it is not humane, and it is not a Universal Truth for the time.

Monsignor was fascinated with my life and told my story to a writer who asked if he could interview me, and I did; later he asked if he could write about my life as an article in *People* magazine. I was not interested; to me, it was nothing but ego. Years later, I saw the movie *Forrest Gump* and said, "This is like my life story!" The motion picture features some similarities in our lives, such as overcoming crippling injuries in my youth, working on a shrimp boat in the Gulf of Mexico, saving lives, and traveling to many countries. Other than that, our stories diverge with different adventures than in the film. I later found out the writer who wanted to put my story in *People* magazine also wrote the movie, *Forrest Gump*.

One day, Monsignor's brother came to visit from Africa. He was a missionary who was dying from cancer. He asked me many questions about life, and we became friends. After he died, Monsignor showed me a letter from his brother that he had written before he died. The note read, "Please take care of Leon because he is a saint on Earth." Ever since then, Monsignor called me Leon the Saint.

After his brother died, Monsignor asked me, "Leon, do you think I will live long?" I said, "You hold no hatred towards anyone and love life. There is no reason you should not live a very long life. Maybe you'll live for 120 years like Moses." After turning 80, Monsignor was asked: "How come you lived so long and still look so good?" He replied, "Saint Leon blessed me."

After I left Our Lady of Malibu Catholic Church, I used to visit the Monsignor and Jack Brady because they were my friends. After the Monsignor turned 85, the church members did not let me, or anybody else see the Monsignor. They said, the reason was, Leon you are corrupting the Monsignor because by debating, he understood that there was a Creator; and for that, the church members ex-communicated the Monsignor. This shows that religious people are very cruel. So, when he was 95 years old, sadly, he drove north to the wine-tasting country, Pismo-Beach, and died in a car accident in 2010. **So, please don't tell me that religion makes you moral!**

When you memorize knowledge you can become the president of the country; however, when you free self-think, in God's reality, you will create a better world for life, and that's what all Saints and humane-beings do.

JACK BRADY

I knew Jack Brady for about thirty-five years. He was my best friend. I worked with him initially in films as a stuntman and an underwater cameraman. One of the stunts I performed was a TV commercial for Timex watches, which ultimately became the longest-running commercial in television history, viewed for over 25 years.

For the commercial, I drove a boat called Espadon, through the waves and flew onto the beach while the watch attached low on the speedboat's bow, "took a licking and kept on ticking," as John Cameron Swayze narrated. While working on the commercial, I met a young man, his wife and two kids. He drove the fastest boat in the world—Miss Budweiser—hitting 225 mph on top of the water. I sat and had drinks with him every day while doing the Timex commercial.

I said to him, "I like to challenge life and have been doing crazy things ever since I was a child. However, what you are doing is suicide; you are going to kill yourself! Life is very precious, don't waste it." He said, "Oh no, I know what I'm doing." When people are going that fast, they have no control and don't know what they are doing. Three months later, I saw him on the news. The boat flipped over and crashed. He was killed immediately. I knew him personally and felt very bad because I could not persuade him with my wisdom to take another path for himself.

I always enjoyed teaching kids how to be a lifeguard, especially Jack Brady's children. They asked me once, "Why is it so important to be a lifeguard?" I said, "Once a lifeguard, always a lifeguard." Today, Mike Brady, Jack's second-oldest, is a lifeguard and a stuntman for television's popular *Baywatch* show and earned a Best Stuntman of the Year award for this work in the movie *A Perfect Storm*. I am very proud of him. Another brother, Kevin Brady, is also a lifeguard and a Park Ranger. The third brother, Johnny, lives in Hawaii, working at the beach!

I watched *Baywatch* on TV one night when one of the lifeguards in the show rescued an older man who had attempted to save someone else. They asked, "Why did you go out there, old man?" He said, "Once a lifeguard, always a lifeguard." I knew Mike Brady had something to do with that phrase in the show.

One day, Jack Brady wanted to learn how to dive and spearfish. So, we went to Santa Monica and bought all the equipment required. Later that week, we went diving in the Pacific Ocean. I tied a rope to some sturdy kelp near the surface so the boat would not get loose; this was routine practice. Jack held onto the spear gun as we swam down together. But the water was so murky that we had to get out. As I looked up, I faintly saw a clearing in the kelp field, so we swam to the surface and exited the water.

Unfortunately, due to the underwater current, we ended up on the other side of the kelp field from where the boat was. I told Jack it would be dangerous to go underwater again because it was too murky below; we might get tangled in kelp. Therefore, it was a better idea to swim 100 yards north to where there was a clearing, so we could cross to the other side of the kelp field and come back to the boat. Jack agreed and we started swimming.

After about 50 yards, Jack said he was getting tired. We were swimming against the wind and rough water. I told him to hold onto my shoulders as I kept on swimming. In the afternoon, the north wind picked up and created wavelets crashing on us. One good-sized wave came down and Jack pushed me under the water to breathe with his head above the wave. I was surprised and swallowed a big gulp of seawater and could not breathe. I was coughing when more and more waves splashed down on us. I thought I was going to drown. I raised my right hand in the air and said mentally, "God! Please help me! I am not going to let my friend drown. I am not going to push him away to save my life." We were both desperate, exhausted, and succumbing to the cold.

Then a crazy thing happened. As I ferociously moved my arms to stay above water, my hand hit a solid object. I turned around and saw our boat only two inches from me. Even today, I don't understand how a boat could disconnect from the kelp, move against the wind for about 50 yards toward us, and end up so close. It was impossible. When I felt the boat on my right side, I hollered to Jack, "Look! Here's the boat! Climb aboard!" Jack exclaimed, "Leon, you're a magician! How did you get the boat here?" I said, "You're the saint. You made it happen. Not me. I had nothing to do with it! I thought we were going to drown."

To this day, I still wonder how our little boat came to us in the opposite direction of the wind and current. I will never say it is a miracle! I just accept that there are things I do not understand yet. Now, I am 82 years old, and I realize that I should have been killed more than 10 times; however, I am still here.

Steve McQueen *Vincent Price* *Roy Ach* *Paul Lynde*

These celebrities lived on Nicolas Beach Road

BUZZ ALDRIN

Astronaut Buzz Aldrin and Leon Halfon

Buzz Aldrin is a scientist-astronaut, a space traveler, the second man who walked on the moon. I can tell you from personal experience that Buzz Aldrin and many other scientists I have met were very gentle, free self-thinker, humane-beings. Now Buzz Aldrin is 93 years old from embracing science, and not embracing the illusion of knowledge.

CHAPTER 49

HALFON-DULA FAMILY HISTORY

My father built furniture carved from wood and marble, utilizing a six-foot-tall diesel motor since there was no electricity on Djerba's island. The engine mounted a spinning pipe to the back of the wall. On the tube, three leather belts could be connected to three different machines. A person could connect one leather belt to the specific machine they needed to use. The motor had to be started by cranking a lever.

MY FATHER, MERO

Wearing a French Police Hat

When I was a little kid, my job was to empty the sawdust in the shop and put it into burlap bags, sold for agricultural use because the island had no soil. The whole island was like one big beach.

When I was only five years old, while I was underneath the saw machine to take and empty the sawdust bin, seven religious Muslim bullies came in with knives on their belts. They were threatening my father about wanting a money bribe for the Muslim religion leader. My father refused. I stayed hidden in the cage under the wood cutting machine. My father did not know I was in the shop. That is the concept of Babylonians creating religion in blind-faith. By stupid people's cruelty, religious leaders receive money for doing nothing, it is still happening today. **That is stupid!**

One of the thugs started shaking the diesel engine that was bolted onto a concrete block. It looked like they were going to destroy the place. My father had been carving a marble piece with a massive hammer. He threw the hammer at the marauder, who was shaking the motor. The hammer struck his head, dropping him to the floor. His head fell ten inches from where I was hiding. I saw his eyes trembling and his head bleeding profusely. The blood poured under the machine, soaking the sawdust. I could feel the warm oozing liquid on my knees. His open eyes fluttered as I watched him stiffen and die. My father almost started an all-out brawl, but fortunately, the French police showed up and the Muslim thugs disappeared. The police hauled the dead body away and nobody went to court. Thank you French police, because this world belongs to everybody, not to god players.

I recall another time when some other fanatic religious thugs were having a protest right in front of my house. They were demanding that everybody who was not Muslim should leave the country. I was on the balcony of the second floor at home and witnessed everything.

My father and his friends walked toward the house as they heard the violent men screaming. My father went toward the wall and pointed at the leader who was preaching loudly and said to him in Arabic, "Hey, you, tough guy! You have a knife. I do not. So why don't you come and kill me or are you afraid?" The thug came running, and screaming, with his knife in the air to stab my father. My dad jumped aside quickly, grabbed him by the shoulder, and ran his head into the wall. He died instantly.

Another furious protester ran toward my father and pulled out the knife from his belt when my dad rushed toward him. He grabbed and twisted the assailant's arm, which forced him to stab himself in the stomach.

These were the three times I saw my father kill someone, but I am pretty sure there were others. My father killed these men because he did not want to be told how to live his life. He knew that these fanatics felt that they had the right to control his life because of their religious beliefs from power as a majority gang. My father needed to defend himself and his family. My father was a free self-thinker powerful man with true grit. The French police liked him very much and they told me so.

When I was five, we lived in a building on the second floor, which had tunnel stairs made of concrete from the Romen builder. I'll never forget when a group of fanatics attacked our house in the middle of the night to kill my whole family. They were screaming and brandishing their guns. My father told us to hide, so the intruders could not find us. I was scared for my life but was also curious and wanted to watch. I could hear the marauders climbing the stairs. Then suddenly, my mother opened the door. My father stepped forward with a bucket of gasoline and poured it into the stairway tunnel dousing the attacking men. Next, he lit a match and threw it at the thugs. The entire stairway burst into flames. The fanatics in the concrete stairway burst on fire and ran away screaming. My father said violent people are animals; they are afraid of fire, and they are not human-beings in wisdom. **This is another example that shows that religion, and tradition in blind-faith, destroy the utopian democracy on the peaceful island of Djerba!**

Four hundred years ago, my dad's great-grandfather was a sailor from Zebulun the lost tribe of the Hebrews. He loved adventure and fought against the Spanish Armada. In one of these battles, he captured a Spanish ship sailing from America to Spain with its cargo hold filled with gold, because stealing from a thief is not stealing, in Hebrew wisdom, and also, they did not approve of hierarchy and became full-fledged pirates based in Tripoli because hierarchy is not by the people, for the people! Everybody nicknamed my great-great-grandfather "Dula," which means "man in charge. He loved serious fighting, and his preferred method was kicking his opponents' legs, shins, or groins. Part of his training included hitting complex objects on his legs to toughen them. He trained his children this way and my father trained me. It is a very effective way to defend yourself.

Dula and his men decided to protect and help the United States financially, with their gold, because he disapproved of world hierarchy. They served under George Washington and his American commanders during the Revolutionary War. Because of his service, the document shown below is over eighty years old and it talks about another document over two hundred years old. It says Dula's family is protected under the aegis of the United States flag. A transcript follows.

The first treaty of peace and friendship was between the United States and Tripoli, which was approved by George Washington explicitly stated: "The government of the United States is not in any sense founded on the Christian religion..." This treaty was negotiated by the American diplomat, Joel Barlow, during the administration of George Washington. Washington read it and approved it, although it was not ratified by the Senate until John Adams had become president. Free self-think to understand why we approve of world hierarchy and religions today. Is this a political movement to get rid of leaders.

U. S. Consulate

This is to certify that Nassim Halfon, son of Maher Halfon, late broker to the United States Consulate at Tripoli of Barbary and grandson of Aron Halfon, the previous broker of the same, came this day to the Consulate and declared under oath that for the last eighty (80) years, his family was under the protection of the United States flag; that every member of it who, during that lapse of time, was born, brought up, married or died, was considered as being an American protégé and that he, having never done anything contrary to that political status, claimed to be entitled to the continuance of that protection. This document was delivered to him to certify that the United States protection was duly continued to him.

Given under my hand and the Consulate's seal, February 1876, the 100th year of the United States of America's Independence.

Signed

U. S. Consulate

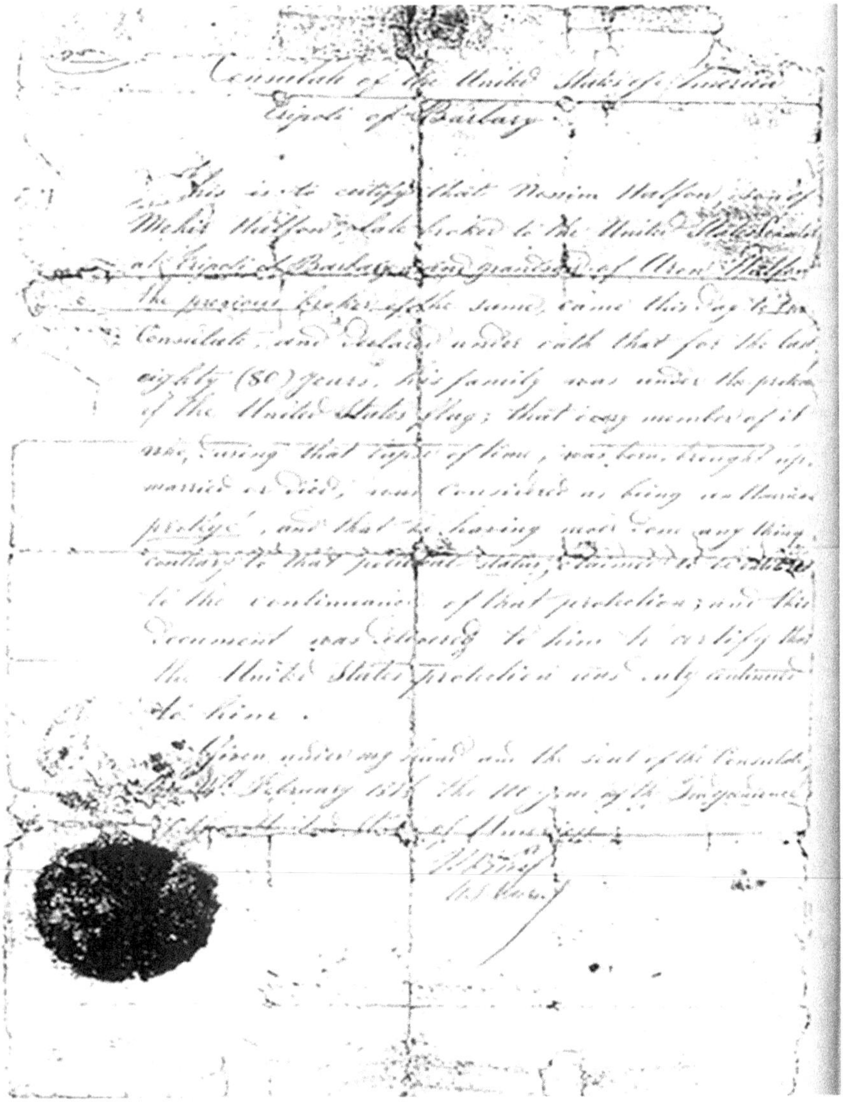

Consulate of the United States of America
Tripoli of Barbary

This is to certify that Nissim Halfon, son of Mekis Halfon, late broker to the United States Consulate at Tripoli of Barbary, and grandson of Abert Halfon the previous broker of the same, came this day to this Consulate, and declared under oath that for the last eighty (80) years, his family was under the protection of the United States flag; that every member of it was, during that lapse of time, was born, brought up, married or died, was considered as being an American protégé, and that he having never done any thing contrary to that political status, claimed to be entitled to the continuance of that protection; and this document was delivered to him to certify that the United States protection was duly continued to him.

Given under my hand and the seal of the Consulate this [illegible] February 1876 the 100 year of the Independence of the United States of America

[illegible]

Consulate of the United States of America Tripoli of Barbary

My father told me, and his father said to him that Dula buried a chest of gold in the middle of Tripoli harbor for future generations, but I am not going there because it is full of religious gangs and too dangerous.

My father's grandfather was named Abe Halfon-Dula. He owned his ship and sailed across the Mediterranean Sea to transport carob beans, a fruit crop **used as money.** A strong storm sank the ship one day and Abe ended up sitting in a large wooden food bowl. In those days, the mariners used a four-foot diameter food bowl to prepare food during their journey. The main meal consisted of a mixture of barley flour, oat flour, almonds, walnuts, dates, olive oil, and brown sugar. To this day, I prepare and eat it myself; I call it pirate food. Abe floated for eight days in the food bowl and landed at an Egyptian harbor on the Nile River. Two free self-thinking humane-beings rescued him. They gave him a hot bath and food and sent him on his way. Abe worked as a repairperson and eventually made it back to Tripoli.

I recall a story about my father's great-grandfather. He was a professional brawler, and his name was Halfon-Dula. In the old days in Tripoli, fighters would participate in tournaments once a year without weapons. The winner made enough money to live comfortably for a year, but some fighters died, and others were seriously injured. When Tripoli became an Italian colony, fighting for prize money was forbidden. Meir Halfon-Dula's new job was to transport money for business transactions. People knew that he carried money and one night when he went to sleep, he was murdered for the money.

Listening to these stories and observing my dad made me want to prove myself. We kept a big black rooster in the workshop's backyard. He made lots of noise and nobody could catch him. My father offered 100 francs to anybody who could catch the rooster. The rooster was three feet tall. I was six years old, and I was going for it to prove myself.

I went to a corner of a deserted building in the backyard where it was shady; I crouched in the corner and waited for the rooster. Sure enough, the big bird approached two feet from me and started scratching, dancing, and challenging me to a fight. He had sharp spurs on his feet, and I watched them closely. I crouched down to make myself small, so the rooster felt he could attack me. As soon he was close enough, I jumped, grabbed his feet, and started screaming for help. The rooster pecked at my hands. By the time help arrived, I was bleeding seriously, but I would not let go.

My father was proud of me, and he said I was a real Halfon-Dula.

My father was a powerful man, but he got sick at seventy-five from a heart condition. When he was young, it was a tradition to eat sweet, barbecued lamb fat. Now we know that is bad for the heart, so we no longer do it. It does not matter if it's tradition, beliefs, wrong, or harmful opinions; it is terrible for you and the world at this stage of human evolution. So please, free self-think, and understand not to do it!

If we all decide not to follow or associate with people who are not humane and choose to hang out with only loving and caring humane people, we will live long and meaningful lives. When you are a free self-thinker in wisdom, you can do just that, so do not let people who are not humane into your life, your country, or your world!

MY BROTHER, YEHUDA

Yehuda was a paratrooper in the Israeli Army. During 1967, in the Arab-Israeli conflict known as the Six-Day War, he and his platoon parachuted into the Sinai Desert. The soldiers eventually began to run out of food and water and had to wait to be rescued. Meanwhile, twenty Muslim combatants arrived with their hands in the air. They surrendered because they had no food or water as well. The Muslims knew that if they surrendered, the Israelis were obligated to share their food and water. However, the Israeli captain told my brother, "We don't have enough provisions for ourselves. We will starve if we have to give some of our water and rations to these Muslims." He ordered my brother to send them away and if they did not go, shoot them.

My brother refused to obey his superior officer's command, infuriating the captain, who was about to discipline him on the spot for ignoring a direct order. In the end, they and the Muslims were rescued by the Israeli Army, and nobody died. Afterward, my brother asked the captain, "Don't you feel better now that you didn't have to kill those enemy soldiers or send them away to die?" This story's moral is that you are also guilty if you obey inhumane orders—no excuses. That is why it is essential to free self-think to do what is right! When people believe in sin, it incorporates unnecessary factors into their choices. When people rely on religious Babylonians (god-players) for forgiveness, they lose their connection to humanity and God. Do not let anyone tell you what to think. Analyze the situation, consider all relevant information, and make what you think is the right choice. Free self-thinking in search of universal-truth is the only way to have confidence.

MY MOTHER, RACHEL

When my father died, all eight remaining children moved out independently and my mother rented a small apartment for herself. One day while she was out, a burglar broke in and stole her valuables. The culprit was caught, and my mother went to the police station to press charges. When she got there, she asked to see the prisoner.

When they brought him out, she noticed he was a Hebrew. My mother told him, "You must have had a terrible experience to go the wrong way and become a thief! Don't you realize you are a Hebrew and Hebrew people are free self-thinkers in reality to be moral and humane—not criminals! It's written in the Constitution of people and God's 10 Commandments; you are supposed to do better." She gave him a hundred dollars and said, "Straighten out your life." Then she walked out without pressing charges.

My mother and father with six children.

PART SIX

LEON'S FREE SELF-THINKING IDEAS

CHAPTER 50

INVENTIONS

All ideas are in the realm of wisdom by asking yourself questions and participating in scientific debate conversations. By searching in your mind to find, from an accurate understanding of reality, you will get confidence by free self-thinking some great ideas for a better world. However, you cannot be creative, from an illusion of knowledge or a blind-faith gambling addiction.

The following inventions are for people who can use these ideas to make their lives and the world better. Please contact me if you are interested in understanding more about any of these ideas.

SOLAR WATER DISTILLATION SYSTEM FROM HUMIDITY

The first invention I would like to share is a distillation system. It takes water from the atmosphere and distills drinking water, using only solar energy. Build a black box, 4 ft. By 8 ft. by 4 inches deep and insulate it. Make a trough all around on top, sloping glass to the drain at about 20$^{\%}$ sloop. Fill the box with sodium and leave it out at night. The container should be open some, so that air can come in. The sodium absorbs the air's humidity to create a small pond about 1" (inch) deep. During the day, you close the box hermetically so that no air can come in or out. On a regular sunny day, it will get 225 degrees in the box. The water then evaporates, hits the glass, and slides into the trough where it's collected outside. It will evaporate about 1/8th inch. You can expect to accumulate about 2 gallons of water every day from each distillation unit built.

If you put 50 units on one roof, you will collect about 100-150 gallons of water. That should supply enough water for survival. You could probably produce more water in the desert where there are hotter temperatures.

THE BEST SPRINKLER SYSTEM

Next invention: a new sprinkler system. Take a faucet that has a rubber gasket that shuts off by spinning the handle. Install a solid Teflon cylinder with a rubber gasket. When you open it, it will open the valve and when you shut it, it will shut the valve. After replacing it, you put the whole thing in the ground. When the soil is dry, the Teflon cylinder shrinks, and it opens the valve and lets the water out. When the ground is wet, the Teflon cylinder expands and shuts off the water. You can adjust how much humidity you need by turning and spinning the valve open or closed. In this way, you get the right amount of moisture in the ground all the time without mechanical equipment, like timers, which often break.

With these two ideas, you could build a house in a factory and move it to the middle of the desert to have water, electricity, and food to sustain yourself. Electricity could come through solar power and windmills.

REMOTE CONTROL DRONE

Next invention: a remote-controlled drone flying saucer for reconnaissance or advertising. Take two oval Mylar pieces, four feet wide by six feet long, and glue them together with a valve. Put a circular aluminum tube, mount the Mylar to an oval aluminum tube, and connect one tube underneath in an oval. Now use the tube as a gasoline container. Underneath, connect a remote-controlled propeller engine with a small camera. Fill the Mylar with helium. It will take the shape of a flying saucer, and it should lift about six pounds. It will be a little bit lighter than the air. The circular tubing around the drone will retain the Mylar in the shape of a flying saucer. With the remote control, you can guide it for surveillance. If somebody shoots it down, it will descend like a parachute and not crash. You can use it to advertise by having it carry flags or logos on it. You can also use it to freak people out if you paint it like an alien spacecraft!

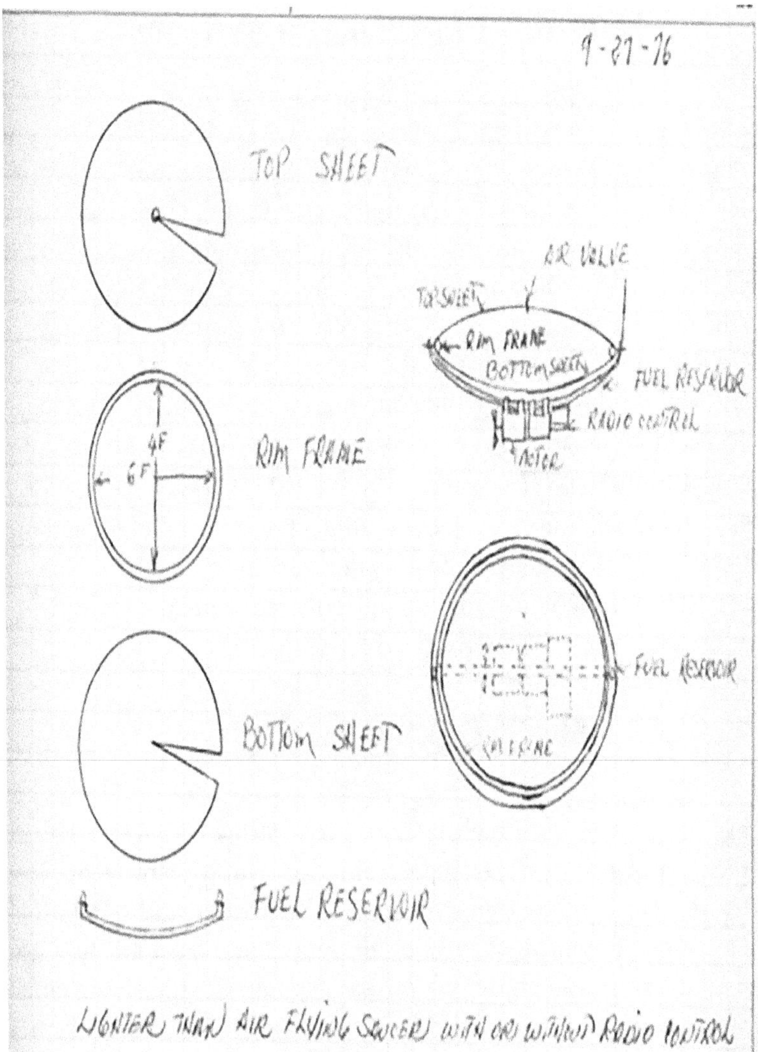

Drone Flying Saucer

INEXPENSIVE SOLAR WATER HEATER

This heats the water that comes in one direction, goes through the box and comes out on the other side. As a result, the water will be hot. This is inexpensive and simple; it works better than other solar water heaters because the box does not get hot, but the black tube warms to create heated water.

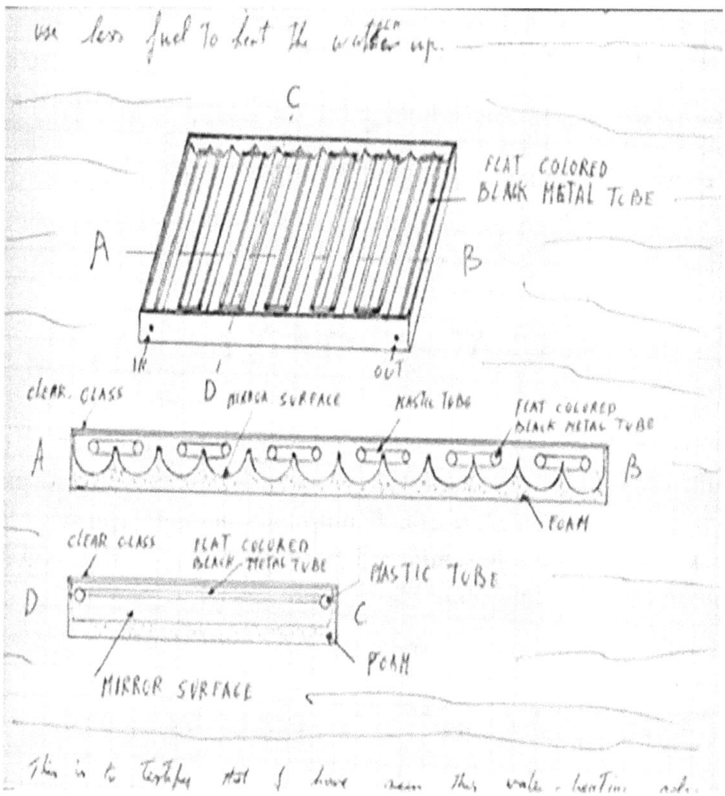

An inexpensive solar water heater. Build a mold that is 4 inches thick by 4 feet wide and 8 feet long. On the inside, every 3 inches, make a parabolic shape along with the box, which you fill up with foam. Glue mirrored plastic or aluminum on the top, gives the inside a mirrored effect. Put a wooden frame around it and drill holes in the wood in the middle of each parabola shape. Install black plastic tubing or brass tubing (paint it with flat black paint) in all the parabola shapes and connect it outside. Put a piece of glass on top. When the sun shines on top of the unit, it will reflect the box's heat into the black tube.

PRODUCING ELECTRICITY WITH NATURE

You can use old ships past their prime—like the Queen Mary in Long Beach, California—and have them float to sea, not too close and not too far from the beaches. These ships weigh over a million tons. One ship going up and down by the swells in the sea will produce enough energy for a whole city. Water pumps will operate a turbine that creates electricity. It will also make the beach calmer and safer for swimmers!

ETHANOL AT FIVE CENTS A GALLON

Companies like Sunkist are willing to deliver ten truckloads of orange pulp a day for free. Put a massive, black container tank in an open space on top of a hill and dump the orange pulp into the tank. The sun's heat on the black tank ferments the pulp and produces ethanol using gravity and distillation. Gravity and solar energy are free!

BUILD A CONCRETE HOLLOW WALL

Strong and inexpensive! Put two pieces of exterior plywood metal mesh together with PVC pipe and wood in the middle. Screw it together and you will have a form to install over a footing. Once installed, pour the concrete into the middle. The form will not separate from the concrete pressure because the tubes hold it together at the proper distance.

CONVERTING TO A NON-SKID BATHTUB

Great for elders! Instead of buying an expensive walk-in tub, this is very simple and inexpensive. Coat the tub with concrete glue and then install a good quality white thin-set. Place 1" square tiles over the thin-set. When all is finished, grout and seal. Nobody will slide in the bathtub and have an accident. It will always stay spotless if you use bleach. It lasts a lifetime. If you need someone to do this, contact a tile installer. Note the picture below from my tub.

REPLENISH THE DEAD SEA IN ISRAEL

The Dead Sea is the lowest spot on Earth. It is drying up at an alarming rate. After all, the River Jordan does not flow into the Dead Sea as in the past because people overuse the fresh water from the river. My idea is to install a giant pipeline from the Mediterranean Sea and pour water into the Dead Sea by gravity. Again, the Dead Sea is the lowest place on earth. On the way, the seawater will pass through an osmosis system, which will separate the seawater into freshwater and salty water, by the energy of gravity.

Fresh water could be used for agriculture, and the salted water will go to the Dead Sea. On the way, it would go through a turbine which would create electricity. The project could produce fresh water and electricity and replenish the Dead Sea. I used to work there, and I know this is important because it's the sea with the most concentrated minerals on the planet. Pilgrims go to bathe in the Dead Sea's minerals to improve their health. This idea will help.

SOLAR AIR-CONDITIONING

This is like a propane refrigerator in a travel trailer. Instead of a propane flame heating a coil, the sun heats the coil using a magnifying glass and a solar cell that guides sunlight to the coil, creating refrigeration. The hotter the sun, the colder the unit will be.

THE SOLAR ENERGIZER NECKLACE

The solar cell is cast in an acrylic pendant designed to focus light onto the solar cell. The left and right halves of the chain are the positive and negative electrical sides. They are separated by the plastic insulator, making sure the current flows into the neck's back where your central nervous system starts. It produces 10% of the electric current of your body to make your nervous system healthier by exercising the nervous system.

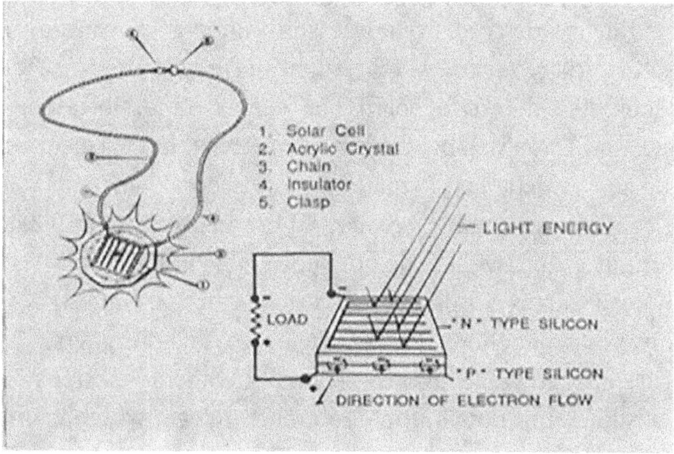

A HUMANE NON-FATAL MOUSETRAP

The little bit of food on the postcard should be very light, such as a smear of peanut butter or a small piece of cheese. The animal falls into the bucket and is not killed and lives to compete in Nature's Laws again. All inventors are free self-thinkers in wisdom. That gives humane-beings the confidence to be creative and make the world a better place. Please free self-think and use debating conversations in reality for a more accurate understanding to find new wisdom that creates new knowledge!

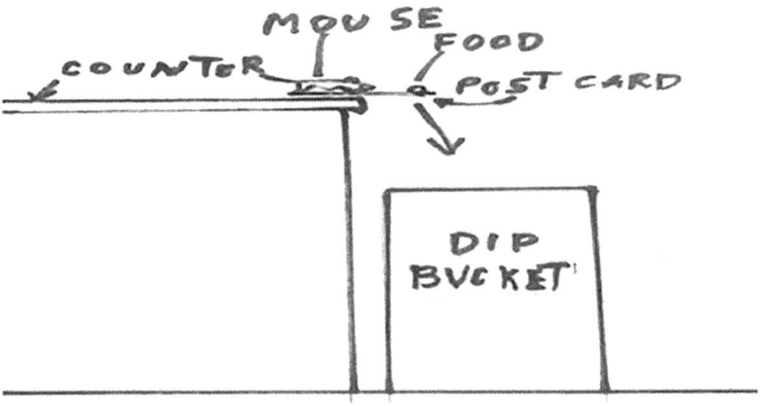

I lived through all of my exciting, sometimes scary life stories, from having confidence derived from free self-thinking by learning from my ancestors' inspiring stories, which taught me how to ask questions. The answers look like a puzzle board that seems impossible; however, by handling myself I was willing to learn from others. In the scientific method of thinking and by debating, I created a more accurate understanding of the spectrum of life magic reality because all the answers are in the questions. Even if you say: "I know that I do not know," this is a valid answer that will ultimately lead to new wisdom. However, if you don't question what you don't know, you will believe Babylonian leaders and preachers. That is why you must ask questions and debate to learn from others in search of universal truth for wisdom**, and not compete to create an ego, which is inhumane!**

CHAPTER 51

FREE SELF-THINKING CREATES WORDS OF WISDOM BY LEON

LIFEGUARD POEM

I liked to sit on the beach. In the evening,
After everyone was gone,
I would get a great feeling.
As it got darker,
Gazing at the glowing, pounding waves,
I would raise my eyes and stare at the horizon.
Then, as the universe meets the ocean, I understand.
All these atoms
Created thinking Adams

A FREE SELF-THINKING PRAYER FOR LIFE TO CARE

This is my moral, humane prayer:
When I am looking at all this beauty out there,
I understand this Energy is in me and everywhere.
Give thanks that I was chosen for life to wear,
And realize it is my responsibility for all life to care.

THE WORD BELIEF

The word belief only means accepting what somebody says. All scams and propaganda are created by believing parasite leaders and scholars that need to be important. The word belief was created by religious fraud for religious dominance over the public to develop a gang for hierarchy. However, this word is often used in people's daily lives because the word belief has been imprinted in the public brain by preachers, often preaching the word belief; but many other words can be used instead. For example, I think, I anticipate, I assume, I estimate, I expect, I fancy, I gather, I imagine, I presume, I reckon, I sense, I suppose, I surmise, I suspect.

However, the word belief creates **no certainty, and no confidence to create**. Non-believers, or (atheists), or criminals, or a religious aggressive inferior gang in a blind-faith gambling addiction, to feel important; and that is the reason society needs, for every 100 people, a lawyer and a big social government. However, if you want to arrive at an accurate understanding in a conversation and are searching for universal truth in scientific reality by debating, you will understand, without a doubt, that there is a creator, once you do. No leader, cruelty, or propaganda is capable of making you lose confidence in yourself or demoralizing you or making you **immoral**. So, people can become free self-thinker conscious humane-beings in a creative democracy. **So please don't tell me that religion lies, creates beliefs, makes people moral.**

If you think this life comes from nothing or evolution, then you are a believer in preachers, who write atheist scientific philosophy; therefore, you are not a free self-thinker, because any fool can memorize scientific knowledge and write what they believe. Because the more you believe, the more you think you know everything, and that makes you feel important. The concept is to free self-think and understand for yourself by learning from others; and the more you understand the reality of science, the more you realize that you don't know much, and that is challenging to excite your mind. However, evolution is evolving life to adapt to the environment, and energy is in everything. It can't be created or destroyed; only life-energy creates life by the Creator in reality of science.

What believers do best is justify and justifying is not an answer or reality, it is just an excuse! However, when you say I believe, you are uncertain, because if you are certain, you say I know; and when you say I know, it doesn't mean I understand what I know, and when you say I understand what I know, it doesn't mean I accurately understand the universal-truth of that time!

Think of life as a clock; all the gears must work correctly to give you the right time. But if the gears don't correlate in your brain from religion, tradition, and blind-faith, you will have to take somebody else's word for the time, (that is believing somebody else) although you cannot know for sure. It will instill false confidence. Consider the power of suggestion, the power of intuition, the power of the mind, hypnosis, the power of nature, the power of love, and the power of life energy. And by scientific method debate, you will create an understanding of these powers for a healthier life.

Free self-thinking in the search for wisdom goes far back to when people started thinking and there is no date known for that because people have always had to think and do things. On the other hand, all beliefs from religion, Buddhism, and atheism can be traced to the date and time when their leaders started their hierarchies and gave their gangs a name and a religious faith from the illusion of knowledge to compete with another team. That is what all corporations do to create more business. Atheism began to appear in Europe and Asia in the sixth century B.C. Philosopher Will Durant found that certain tribes in Africa and all animals are atheists.

The downfall of religion is the ego of its leaders, and the failure of its followers is ignorance and laziness brought about by the absence of thinking for oneself in a debate! Where there is life, there must be something that makes life possible. People with high IQs are tired of being scared by cruel violent religious god players and their illusionary God. They choose to become atheists because they understand that fearing God is wrong and having high IQs is having high **knowledge, not scientific wisdom**. When we learn in books or movies about enchanted beliefs like those of the ancient Egyptians, Greeks, or Romans, we laugh because we cannot understand how anybody could believe all this ignorance. It seems to be the truth, just like today's religion, but all these imaginary beliefs are illusions of knowledge. In the future, people will look back at the beliefs of today and laugh. Belief is like knowing there is an address for God somewhere. Followers believe in blind-faith leadership. Leaders give wrong directions because they lose themselves because of their egos. They clash with others in hatred and wars to prove who is right. Believers think their path is the one natural way to go, but that is an illusion. This is why atheists came to think that religious atrocities are wrong. They are tired of being afraid of God, they are in the illusion of themselves as gods. In reality, they have all died or will die someday; therefore, they are not gods. You wouldn't want them in charge of your life or any lives.

If you think there is no Santa Claus and no man in the sky named God, that doesn't make you an atheist. On the contrary, it makes you a realistic free self-thinker in wisdom because this is factual in science. You are an atheist only if you believe there is nothing in the universe that made life possible. If so, you belong to Atheism's gang; you are a believer in an illusory philosophy that **is not a threat to the religious monopoly**.

During a debate conversation, someone told me he was neither Protestant nor Catholic, but he was a Christian. I asked him, "Why do you think you are Christian?" He replied, "I like the humane things that Jesus said and did not accept some of the other church teachings. I think some church leaders have diverted from the legacy of Jesus Christ." My response: "You are neither Protestant, nor Catholic, nor Christian. You are a free self-thinker in wisdom because of your search for what is right and wrong.

However, Mason's Democracy by free self-thinkers in the search for scientific reality is a defiant threat to the religious monopoly of hierarchy. Since the beginning of history, gangs of believers were the majority; and that is the reason they tortured and killed free self-thinkers. They wanted to alter truth and knowledge to control the thinking of their followers. The American Constitution separates religious beliefs from the government. Since then, free self-thinkers in search of wisdom have at least enjoyed some legitimacy and humane protection. **As a result, they produced more factual knowledge in the past 100 years than in the last two million years! Because the free self-thinkers were protected. Imagine that!"**

THE WORD OPINION

The opinion is a stinker, everybody has one, the word opinion doesn't create a roadmap for continuous thinking**, the word to use instead, this is what I am thinking!**

For example, India's subcontinent has freedom of religion tradition, and opinions, in the widest variety of beliefs than any other country in the world and started the Buddhist religion. Still, unfortunately, these omnipresent religions keep people ignorantly idle from believing in illusion. India has the second-largest population on earth, over one and a half billion people and leads the world with the most poverty, starvation, and disease. It is the only place where the rat-born Buddhist plague still exists. Moreover, in the recent past, India was where smallpox killed millions of people every year. To stop that deadly toll, it took one free-self-thinking, hippy son of a Hebrew *immigrant who set up a program to vaccinate millions of people a year, which practically eradicated smallpox.*

Today, smallpox is virtually non-existent worldwide. However, before that achievement, more than one billion Indian citizens and millions of educated doctors, professors, and scholars in their country were unable to eliminate smallpox because all of them were worshippers and followers, thanks to "freedom of religions to create gangs," which gave them the right to remain in illusion of knowledge in blind-faith. They have the illusion of knowledge and knowledge, but no free self-thinker for wisdom to understand reality. No one can change anyone in life, no matter how good the intention or wise approach. However, in a debating conversation in reality, you can let others say what they believe. Then they ask questions to reveal the contradictions in their thinking in the reality of science, to ask themselves important questions and eventually arrive at an accurate understanding of God in reality. Encourage them to do the reasoning in the search of universal-truth and in a debate, create a more accurate understanding.

In the medical field, some doctors use computers not so much to improve knowledge, but rather to record patient information simply. The majority of the medical profession and its powerful affiliated industries strive to discredit other healing practices to look good in followers' eyes, for more power, so they will feel good about themselves. Successful alternative treatments and nature-based, holistic approaches to healthiness should be encouraged. Of course, this presents an existential threat to the medical monopoly, which is why most of these outside-the-box healing processes have been opposed and marginalized for years by the medical establishment aligned with health insurance companies. Doctors should be humane-beings and focus on healing, not competing to make more money.

RELIGION IS A PLACEBO

In good comparison to religion is smoking cigarettes. When cigarettes first came out, cigarette salespeople were in propaganda campaigns advertising lies, that cigarettes would cure practically everything. When the factual reality came out, smoking cigarettes created particular cancer types, and started fires that destroyed lives, homes, and businesses. The big tobacco companies and their advertisers and salespeople glamorized smoking, praised its benefits, and lied to the public like religious preachers, leaders, and hierarchy for decades until the truth about tobacco dangers finally came out. All the cigarette companies' leaders wanted was money and comfort, so they did not want to reveal death's fatal universal truth by smoking. They knew smoking was dangerous and deadly, and is an addiction that is hard to cure, but they didn't care. It took years before the cigarette industry was found guilty of hiding the truth from consumers and was forced to pay vast sums of reparation funds. But big Tobacco's monetary restitution is not enough because they knowingly killed people, profited from it, and got away with murder.

The tobacco industry leaders who knew about these dangers need jail time. And so do the religious leaders preaching an untruth propaganda scam, that is creating in blind-faith, an addiction that is hard to cure, which they knew the truth all along, because they are atheist leaders, that need jail time too, they created a dangerous world that killed people and profited from it. Restitution is paying bribes. Bribery is the worst crime because it makes all crimes possible. Humane laws are designed to stop someone from damaging others, such as selling cigarettes or preaching untruths by propaganda lies. They knowingly killed people, profited from it, and got away with murder and destruction of the environment; all that is doubly wrong and should be punished! But they don't get penalized because gangs run this world and gangs are run by ignorant hierarchy, (the United Nations) to be in charge of the money, not by the representatives of the free self-thinking humane-being in a democracy. That will make the world joyful and safer.

When people can buy a seat in heaven or pay restitution (bribery) for a crime or pay to obtain power in government, instead po facing punishment. Society is divided into the privileged criminal's royalty and the working populace. This is not a government by the people, for the people. This is corrupt gang-hierarchy favoritism and privileges; and this creates two groups. It makes a world that is not united and in times of trouble, the people pay the price; and without democracy, the people stop being creative. However, people's creativity creates all civilizations, and without creativity to solve problems, all civilizations by hierarchies will be doomed; it has happened before because it does not work. That is why the Government must be by and for the people in a scientific method of free self-thinking by debating in a democracy. It is the only way that works, because People who believe the Bible, the Koran, or Buddha, or the Tora, or any religious book have all the answers, are misguided, and are called "square" because they are stuck in a square book. People who think the dictionary and computers have all the answers are also misguided and "square."

It is comforting for people to believe that the universal truth can be easily found in a square book. Because all they believe and want to know is found in that materialistic brain, by believing in blind-faith addiction. When a conversation consists of what is contained in their box, they become an agreeable Dr. Jekyll. Once the conversation goes outside that square box in a debate for a more accurate understanding of scientific-reality, they become the monstrous Mr. Hyde, mentally limiting their horizons to only what is in their box **by competing ego.** They don't humble themselves to learn from others, to understand the reality of life, because they give up their mind free will, by believing in a blind-faith that is a gambling addiction that is hard to cure, that creates obstacles in the brain to reality!

Greed is the enemy of wisdom. A representative of the people should focus on making the government function better for all the people, not just for the regime, and by assisting the needy, providing wisdom to govern the country and not be divided by religious leaders sectarianism. And political corruption corporate greed should not be a government full of leaders controlling everything by creating laws to benefit the powerful and political "gangs" that essentially run the country. Lobbying that represents an influential leader should not be allowed in government. Instead, they should have a debate to arrive at a more accurate understanding that this country and the world should be by the people, for the people, not by the gangs, for the gangs, because the greatest minds that created all scientific knowledge have the freedom to free self-think to disagree with making a better world. That is the reason all men and women need **to have the freedom to free self-think and disagree in a debate to create an accurate understanding of reality.**

Here's another example of free self-thinking in search of reality, not in illusion: When Galileo was 18 years old, he went to church with his friends. While everybody was praying and listening to the priest, that created a blind-faith, Galileo watched a candle hanging from the ceiling on a swinging rope, where there was no wind, which created a pendulum motion. Galileo measured the time of the pendulum swings and found that every swing kept accurate time. That sparked the idea in Galileo for the pendulum clock. It was more accurate than the sundial and it is still used today. **Religion tortured Galileo because they believed he was lying; however, all religions' leaders are liars!**

Galileo was not interested in believing; he was interested in understanding scientific reality to find the wisdom to create new knowledge. Unfortunately, because he was a minority and not a believer in a traditional obedient follower, he was tortured by religious leaders to stop revealing God in reality by being a free self-thinker. Some religions are still torturing and killing people by calling them liars today, for not being obedient believers!

Galileo was a free self-thinker, not an obedient believer. He worked to end Ptolemy's geocentric model of the cosmos, which said all the heavens and stars revolved around the Earth, and the Earth was flat. That model made people feel important because they believed from their religion that they were the center of the universe, and they were born humans. This earth-first model was a standing conclusion for hundreds of years until Galileo revealed the universal truth about how our solar system works. Even in the face of imprisonment and torture because the Catholic hierarchy said, he was lying, he pursued his path, using his mind in a scientific method of thinking. He created the new scientific knowledge that unites all of life. Because when you free self-think, you will understand what is right, and you will do what is right even in the face of torture! What if Galileo and all other great free self-thinkers in science obeyed the hierarchy's religion and tradition**? They would have been just like most folks who follow like sheep. Then, where would we be today? Preaching is propaganda!**

Hierarchy deals, for example: **I will tolerate your belief, and you will tolerate my belief. We can choose peace among our religions. We agree to disagree. While these are grand ideas, history has shown that these concessions never work out because they are illusionary. These peace agreements last only a short time because followers will remain ignorant, under their god player, and eventually stop tolerating other people's beliefs.**

RELIGIOUS LEADERS ARE ATHEISTS

All great civilizations first evolved from people's wisdom. Later, the atheist Babylonian hierarchy hijacked wisdom and replaced it with religion. Today science in a democracy is taking over the world, and religious atheists leaders escape to communist atheists Rasha, and leave their country as a no man land, full of ignorant people and no leader, today there is no need of leaders and no shepherds of people who create gangs. Read the American Constitution! **People can exercise their right to choose representatives, from the people, by the people, for the people, to be creative in God reality of science**.

"Blessed are the poor" means "Give leaders your money." "Blessed are the meek" means "Follow and obey." "Blessed are the humble" means "Do not question authority so hierarchy can remain in control." "Blessed are the hungry" means "We get fat while you starve." "Blessed are the merciful" means "If you catch priests or leaders doing something wrong, don't put them in jail or hold them accountable for their bad behaviors." "Blessed are pure of heart" means "Switch off your brain and mind, use your blind-faith." Finally, "Blessed are the timid" means "Give us your power and become our slaves."

Reality, science, wisdom, and universal truth for the time—have nothing to do with religion or philosophy or opinion or illusion or tradition or blind-faith or inferior people or gangs or leaders!

When people operate under ideas that sound too good to be true, they end up becoming shortsighted. Something too good to be true is anything that offers a payout that is unequal to the work required. Some believe that an inhumane person can continue to misbehave so long as he or she repents. One has to ask or pay off religious leaders for forgiveness to enjoy an afterlife equal to Mother Teresa's. This gives power and importance to money from religious leaders but does not add up to an accurate understanding **and makes money into God.** I recommend seeing the movie "The Purple Rose," starring Sean Connery, about how religious leaders created a book of propaganda to stay in power over their believers. The movie is based on a true story in history about how the Vatican created an educational book for the hierarchy to know how to maintain control and stay in power by preaching propaganda (scams) to create a gambling addiction **from lies,** for the masses that want to believe that they are winners!

Followers often deny factual reality because the truth is painful when it contradicts their beliefs. But when you hide from the truth reality due to cognitive dissonance, it will become more painful in the future and pain is necessary to know where the problem is. Jesus did what he did because the truth created pain to see human beings tortured and killed by greedy atheist Babylonian hierarchy to protect the basket of money for more power.

To be a free self-thinker, you must be brave enough to stand up for what you know, and brave enough to admit what you don't know. When you are a religious believer, you are unwilling to admit that there are things that you don't know. When you say, "I believe," it sounds like you know everything. However, in God in reality of science, nobody will ever know or understand everything, because life is a journey of inspiring challenges!

When I ask for directions, some people answer, "I believe it is that way or this way," and my response is, "I didn't ask for your blind belief. I want to know if you know where it is because I don't want to go on a wild goose chase!" Just say you don't know; is that so hard? This is why people are on a wild goose chase worldwide because believers won't admit they don't know. If you are not sure, the correct answer is to say, "I am not sure, but I think it is there." It is all uncertain until proven in factual knowledgeable **reality of science. A brain is a terrible thing to waste, clogging it up with unrealistic illusions to create blockage in your brain to reality!**

Some people claim that advanced space aliens-built pyramids and other impressive structures throughout history because they are believers and not thinkers. They don't give enough credit to humane-being wisdom by free self-thinking to become brilliant **because believers in illusion can't believe that you can exist without believing to have the confidence from understanding to be creatively brilliant!**

The afterlife is a commodity that can be sold. It costs nothing to religious leaders, so everything is profitable. It is the best business in the world. With a profit margin of 100% and no need to work, religious structures are among the finest buildings on the planet and rest upon the most expensive real estate. With a market like this, it is no wonder why there are several different denominations within most religions. Everybody wants in on the action. The power that believers can provide. When you are a religious leader, an obedient follower will fight your wars, pay your bills, and raise your children to offer you the same. This is wrong. It gives incentives for people who seek power over others and don't want to work.

Here is an example of what has happened in the United States. In 1830, the Book of Mormons was published. Brigham Young (nicknamed "the American Moses") led his followers to the "promised land" of Provo, Utah, where many schools and churches were founded under his name. Brigham Young sounds like "Bring Them Young," which may explain how he could attract many young people into his flock and why he approved polygamy for many wives to bring more children into his religion. One cannot help but notice how the subtle propaganda of a name like "Bring Them Young" reinforces its mission to captivate the young and the impressionable.

The Mormon founders knew they would not have succeeded in getting followers from other religions because people are brainwashed by childish fairytales not to convert. The Mormons understood that if they wanted their religion to succeed, they would have to preach to gullible children about their illusions. They have to overpopulate to give leadership power. Mormons run their religion like all businesses. They invested in future markets and took advantage of the naiveté and gullibility of children to create followers! Religions' bloody past is evidence that it is taking people and the world in the wrong direction. Its fanaticism can conjure phony reasons for atrocities such as torture and genocide. In recent history, we have seen genocide in Syria, Afghanistan, Iraq, Darfur, Sudan, Nazi Germany, Russia, and China. It has been going on for thousands of years and it is still going on today! It is all about power in cruel inhumane leaders', by scaring people to accept, that their belief is actual wisdom, and other religions are not! because all religious blind-faith is a stupid illusion.

My friend moved to Mumbai, India, and built a lovely house. Due to the lack of bathrooms, many people from the city urinate on their house walls. The man was distraught since it smelled so bad. He decided to build a wall around his home. After it was built, the people started urinating on his wall and that started to smell bad too; he asked me what he should do to solve his problem. I said, "Paint pictures of religious Gods on all the walls." The man did just that and, ever since, no one urinated on his wall again because no religious person would urinate on pictures of gods—and that is wisdom for the freedom of religious countries. Stop praying and expecting God to fix everything! Free self-think, and by debating to understand what is right, in God's reality, you will do what is right, even in the face of torture.

The Shiites don't believe that Sunnis are Muslim, and the Sunnis don't believe the Shiites are Muslim. Muslims don't believe the Jews are the chosen people. The Jews don't believe Jesus is the Messiah. The Protestants don't believe the Pope is the leader of Christianity. Religious leaders are salespeople who sell illusions.

I know by free self-thinking that no religions or traditions are necessary now in this stage of evolution. Stop bickering and move beyond the age of belief, tradition, aggressive competition, hatred, and war. If a group said to me, “Be religious or we will kill you!” I will pretend to be a religious believer**, and that is what the Hebrew people free self-thinkers did by becoming religious Jews**.

THE REPRESENTATIVE OF THE CREATIVE COUNTRIES OF THE WORLD IS PLAYING A GAME OF CHESS FOR WORLD DOMINATION

The game is simple. First, use your free self-thinker realistic creative people that were created by the country of the people for the people, from the people, to create the most powerful weapons; then that government is willing to sacrifice some of its peons to create a reason for war, for world dominance. It is like taking candy from a baby, because the rest of the third world hierarchies created countries of sheep-people, by killing in the past, and are still killing today, their free self-thinkers as a norm, to protect their privileges. These wars hopefully are created to get rid of parasite Babylonian hierarchy, which imposes their illusionary religion as reality on their people, and replaces them with a representative for the people, to protect the free self-thinkers and gives them the freedom to debate in scientific God reality to evolve into a creative humane-being**, because it is okay to kill sheep that overgraze.** To create creative free self-thinkers for a safer world, it is working, as we are seeing, many people today are creative free self-thinking humane-beings in the rest of the world countries; and we're obedient believers in blind-faith because most free self-thinkers are protected by Internet understanding of word safety. However, an illusionary religious hierarchy, the united-nation, still runs the world, today!

Leadership leads a flock of sheep, for their privileges.
A flock of sheep people create Leadership.
Free self-thinking people, create a democratic creative safe world, in God's reality of science, to unite all!

Only by getting involved will you create a more accurate understanding of reality and become one of the greatest humane-beings of the future to make a better world for life!

CHAPTER 52

BETRAYAL AND OTHER WRONGDOINGS

I was 77 years old, a general contractor with a B license. I was hired in a verbal contract by my brother to renovate his second house. I was requested to sell my home and move with all my belongings, tools, and material to fix, and manage the place, and buy it, to be closer to my brother and to stay permanently, because we are getting older. However, after fixing the second house for more than two years, and paying $3000 a month, plus I pay for the water, power, and gas, while waiting for my home to be sold. Now that my home is sold, the value of my brother's house is up too because of my improvements, and the value of all homes are up. So, my brother's wife said, she did not agree with her husband's Verbal agreement, and she is selling the home and gave me 60 days' notice, without compensating me for my work. My brother's wife, after 60 days, on the day I was in the hospital, she put 10% of my belongings in storage, trashed and sold the rest.

I WAS BETRAYED BY MY BROTHER'S WIFE FOR AMERICAN MONEY, AND BY BRIBING MY BOAT PARTNER; IRONICLY, HIS NAME IS CRIES!

When I came back from the hospital, I did not have a place to stay. **As I realize now, at 79 years old, and having been told by my brother's religious wife and their religious friends, that they want me to stop writing and stop lying, because I did not know, and did not memorize a chapter in their Babylonian traditional religious book. Religious people believe that their religion, tradition, and money give them the right to destroy lives and the environment, if not stopped by a majority of free self-thinkers' accurate understanding of reality**......

So, please don't tell me that religion, makes you moral!

To my surprise my older brother came to visit last month, and said, you are not 81 years old, because your younger sister is 81, and I am 87

so you have to be 83 years old. However, I was born at home with no birth certificate, and on my driver's license I am 81, end of story.

The enemy of people that are liars and parasite Babylonian leaders, are the people that search for universal truth. and that is science.

So please be a free self-thinker in a scientific method of debating to increase your accurate understanding of your significance. In search of universal truth, by debating in God the reality of science, for more accurate understanding. Silence= to nothing, or 0, and opinion is a stinker, everybody has one.

If you are having a conversation with dangerous religious believers in blind-faith. Silence= to gold.

If you are having an opinionated, atheist, religious, or traditional, in a debate conversation, you are wasting your time!

Freedom of religion, and traditions, will bring all the diseases and political problems that exist in India to America. It already started to happen… homeless people are dying on the street, leprosy now exists in America, and the caste system, which separates people into categories that is created to increase and protect leaders' (the United Nations) privileges, now exists in America, because religion and traditions that keep people idle, and Creative democracy, cannot work together. Is this what you want in the world?

The pen that writes God's scientific reality for accurate understanding to create the new wisdom to unite all people and get rid of god-players that create competing religions, traditions, and wars, that keep people idle is the God-realistic mightiest weapon of all.

All general saying is wrong, including the general saying in this book, because Wisdom is in the understanding of meaning, not in memorizing words!

I am similar to California's 'Smoky the Bear,' which stomps on smoke to prevent forest fires. I stomp on hate to prevent war.

Sometimes psychiatrists avoid debate conversations because they think they are always right and often their patients do. When people spend their upbringing competing and being told what to do, they think that a meaningful debate for wisdom sounds like **jibber-jabber, because the Hebrew people that jibber-jabber live in the island of Djerba.** They understand the word "knowledge," but they also believe that "there is no meaning for the word wisdom. They think knowledge is wisdom, but wisdom creates new knowledge. Because they use words, not meaning, preachers make people believe that great people who have wisdom are prodigies or chosen by God. It is a hierarchical scam! By preaching and passing laws to create more power for inhumane leaders,' because they are not by the people for the people, and it is against Moses' law between God and people, which is the 10 Commandments that created free self-thinkers, humane-being Democratic creative Hebrews, and Democratic creative Americans!

Free self-thinkers searching for universal truth in God in reality of science do not mind being questioned, and don't create propaganda because, they are searching for the universal truth for that time, and people that search for truth, will only tell you the truth, for a terrific creative safer world!

Traditional believers, in blind-faith religion created by leaders, do not like being challenged. And they must create hatred propaganda by cruelty to keep scared lazy obedient followers from knowing the truth!

If you don't want this to happen again to you or other humane-beings, you will survive because of your love of life, and from an accurate understanding of the Creator of life Energy in scientific reality.

CONTACT INFORMATION

323-449-2325
Leonhalfon777@ gmail.com

HALFONBOOKS.COM & HALFONBOATS.COM

As a humane-being, I accept free self-thinkers to debate this book to create a more accurate understanding, and to be better today than yesterday. You only learn from people who disagree; you don't learn from people who agree, because you already know that, and you don't need a preacher.

The end of this book has empty pages, so the readers can comment on what aspects they disagree with for other readers to create a more accurate understanding scientifically, the Creator the Energy of life.

Thank you to Yvonne Rose, Director of Quality Press Book Packaging Company for consulting and guiding me on publishing my books.

Visit: WWW.QUALITYPRESS.INFO or email: amberbks@aol.com

NOTES:

NOTES:

NOTES:

NOTES:

NOTES:

NOTES:

www.ingramcontent.com/pod-product-compliance
Lightning Source LLC
LaVergne TN
LVHW052257200926
848481LV00002B/393

Contents

About Silosa Consulting Group (SCG) 7

About ISO 9

MDSAP 11

Regulatory Requirements 13

Definitions 14

MDSAP Audit 17

Quality Manual as a POSTER 20

MDSAP ISO 13485 23

MDSAP Audit Checklist 37

Advice for manufacturers 113

ISO 13485 certification costs 113

Bibliography 115

Online Certification 116

Other Publications 118

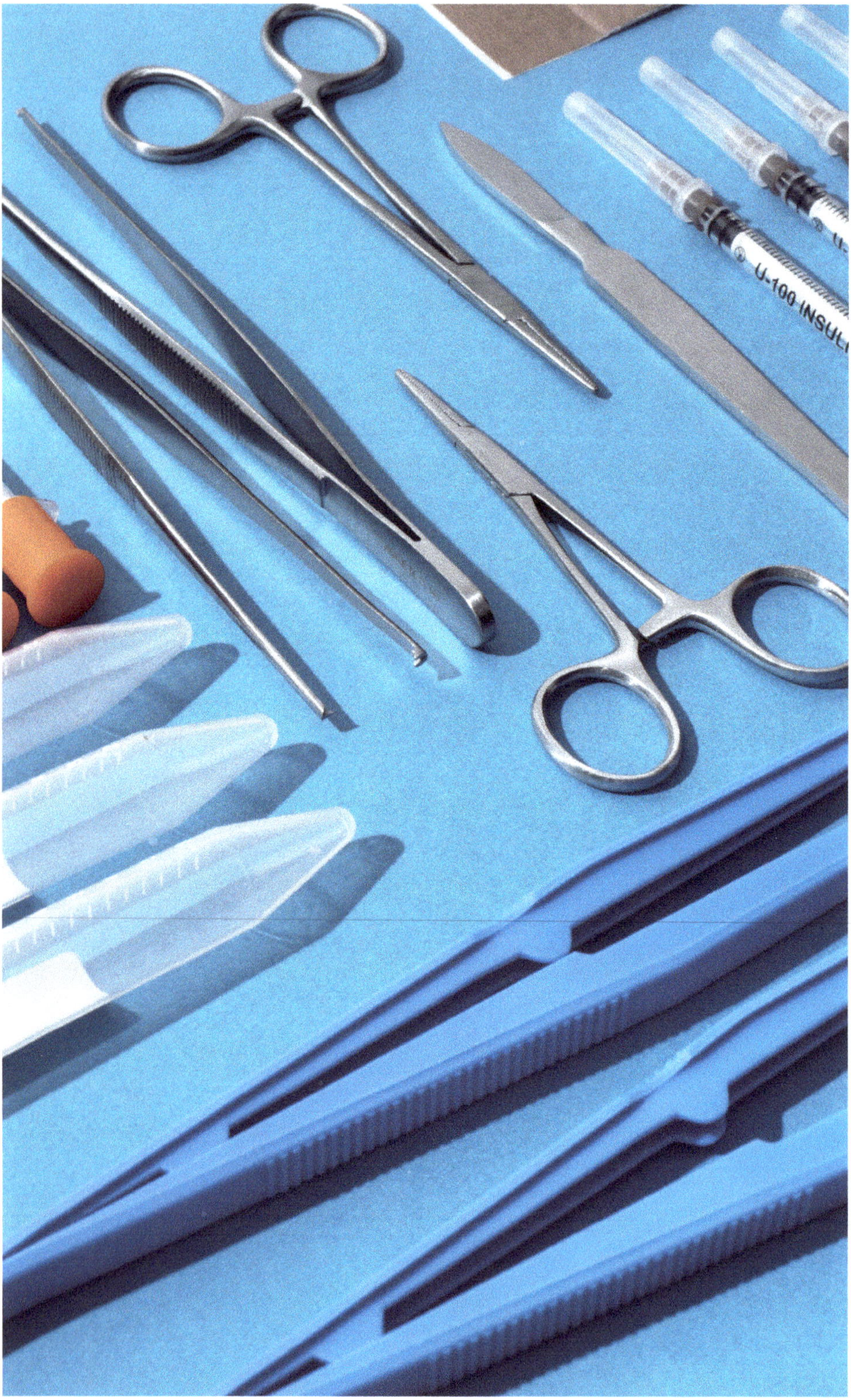
U-100

This book is dedicated to my first manager,

Mir Ahmad Sadat,

a Distinguished man of ISIRI, who has been my supporter and has encouraged me greatly and believed genuinely in my ability with honesty and self-confidence.

SILOSA Consulting Group (SCG)

Silosa Consulting Group (SCG) was established to provide outstanding consulting services of management system standards to individuals, groups, companies, and organizations all over the globe.
SCG is publishing an "EASY ISO" book series related to International Management System Standards to increase public knowledge in implementing these systems over their organizations.
SCG book publishing services and distribution services are connected to over 39,000 booksellers worldwide, including Apple, Amazon, Barnes & Noble, Indigo, Google Play Books, and many more. We focus on quality, environmental, food safety and other management system standards.

SCG has enough experiences to help create new and effective programmes in different countries all over the world. For more detail, visit our
website : http://silosa.ca and/or send your enquiery to the following email:

info@silosa.ca

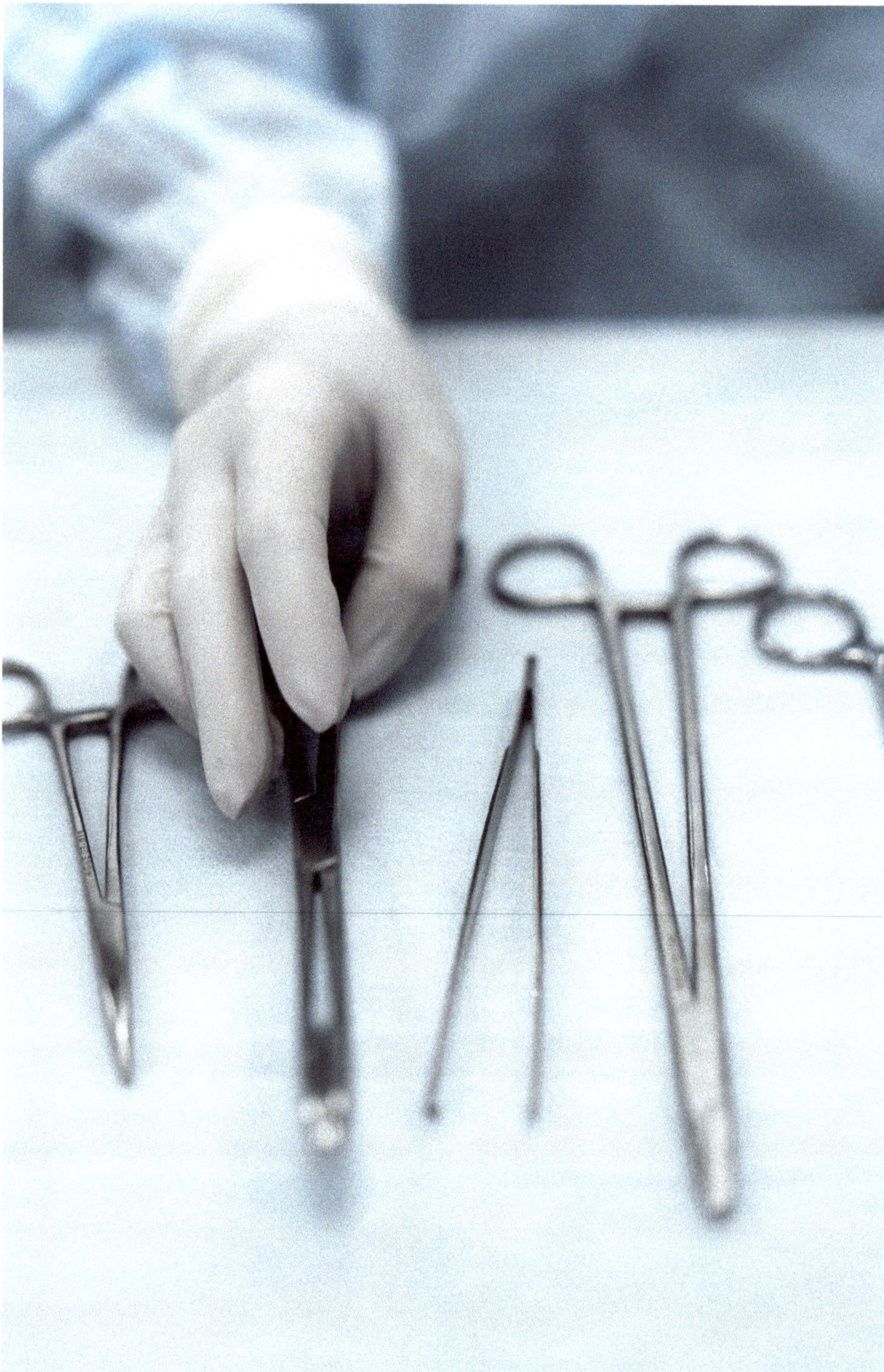

CHAPTER 1

About ISO

The International Organization for Standardization is an independent, non-governmental organization, the members of which are the standards organizations of the 165 member countries. It is the world's largest developer of voluntary international standards and it facilitates world trade by providing common standards among nations. More than twenty thousand standards have been set, covering everything from manufactured products and technology to food safety, agriculture, and healthcare.

Use of the standards aids in the creation of products and services that are safe, reliable, and of good quality. The standards help businesses increase productivity while minimizing errors and waste. By enabling products from different markets to be directly compared, they facilitate companies in entering new markets and assist in the development of global trade on a fair basis. The standards also serve to safeguard consumers and the end-users of products and services, ensuring that certified products conform to the minimum standards set internationally.

History

The organization began in the 1920s as the International Federation of the National Standardizing Associations (ISA). It was suspended in 1942 during World War II, but after the war ISA was approached by the recently formed United Nations Standards Coordinating Committee (UNSCC) with a proposal to form a new global standards body. In October 1946, ISA and UNSCC delegates from 25 countries met in London and agreed to join forces to create the new International Organization for Standardization. The new organization officially began operations in February 1947.
More information can be obtained :

www.ISO.org

Geneva, Switzerlan (Headquarter of ISO)

CHAPTER 2

MDSAP

MDSAP The Medical Device Single Audit Program is the single audit program that covers the regulations of Australia, Brazil, Canada, Japan and USA. These regulations use the MDSAP to verify compliance to the quality systems of their national regulations and ISO 13485. The intention of the Medical Device Single Audit Program (MDSAP) is to allow competent auditors from MDSAP recognized Auditing Organizations (AOs) to conduct a single audit of a medical device organization's quality management system that will satisfy the requirements of the medical device regulatory authorities participating in the MDSAP program.
Audits performed under the MDSAP program will be process based, focusing on several defined processes, a defined method for linking those processes, and built on a foundation of requirements for risk management.

ISO 13485 MDSAP are two different programs with similar requirements but they do not duplicate each other.
MDSAP has the more stringent requirements of the two and companies that are already certified to ISO 13485 will see an increase in the number of audit days once they seek certification to MDSAP.
The design of the Medical Device Single Audit Program (MDSAP) audit process is to ensure a single audit will provide efficient yet thorough coverage of regulatory requirements.
These requirements include; Medical devices – Quality management systems – Requirements for regulatory purposes (ISO 13485:2016), the Quality Management System requirements of the Conformity Assessment Procedures of the Australian Therapeutic

Goods (Medical Devices) Regulations (TG(MD)R Sch3), the Brazilian Good Manufacturing Practices (RDC ANVISA 16/2013), the Japanese Ordinance on Standards for Manufacturing Control and Quality Control of Medical Devices and In Vitro Diagnostic Reagents (MHLW Ministerial Ordinance No. 169), the Quality System Regulation (21 CFR Part 820), and specific requirements of the medical device regulatory authorities participating in the MDSAP program.

What is the meaning of ISO 13485 certified?

ISO certification is a seal of approval from a third party body that a company runs to one of the international standards developed and published by the International Organization for Standardization (ISO). ... ISO 13485 helps put your customers first.

Is ISO 13485 certification worth it?

ISO 13485 is important to designers, manufacturers, and distributors of medical devices. In addition, suppliers and service providers can enhance an organization's marketability as more and more manufacturers require certification in order to do business with a vendor.

MDSAP AUDIT:

The design and development of the MDSAP audit sequence allows a logical, focused and efficient conduct of an audit. The MDSAP audit sequence follows a process approach and has four primary processes - Management process, Measurement, Analysis and Improvement process, Design and Development process and a Production and Service Controls process with links to the supporting process for Purchasing.

The definition of each process includes a purpose and an outcome that are indicators of process performance. Each participating Regulatory Authority expects that risk management to be the foundation for the five processes that are the requirements of a quality management system for medical device organizations.

The MDSAP audit process has two additional supporting processes: Device Marketing Authorization and Facility Registration and Medical Device Adverse Events and Advisory Notices Reporting. These processes are necessary to fulfill specific requirements of the participating MDSAP regulatory authorities.

Regulatory Requirements

Country	Regulatory	MDSAP
Australia	Conformity Assessment Procedures of the Australian Therapeutic Goods (Medical Devices) Regulations (TG(MD)R Sch3)	Vo.1
Brazil	The Brazilian Good Manufacturing Practices (RDC ANVISA 2013/16)	Vol.2
Canada	Medical Devices Regulations SOR/282-98 and specific requirements of the medical device regulatory authorities	Vol.3
Japan	The Japanese Ordinance on Standards for Manufacturing Control and Quality Control of Medical Devices and In Vitro Diagnostic Reagents (MHLW Ministerial Ordinance No. 169),	Vol.4
USA	FDA Quality System Regulation (21 CFR Part 820)	Vol.5

Definitions

ISO 13485 specifies requirements for a quality management system where an organization needs to demonstrate its ability to provide medical devices and related services that consistently meet customer and applicable regulatory requirements. Such organizations can be involved in one or more stages of the life-cycle, including design and development, production, storage and distribution, installation, or servicing of a medical device and design and development or provision of associated activities (e.g. technical support). ISO 13485 can also be used by suppliers or external parties that provide product, including quality management system-related services to such organizations.

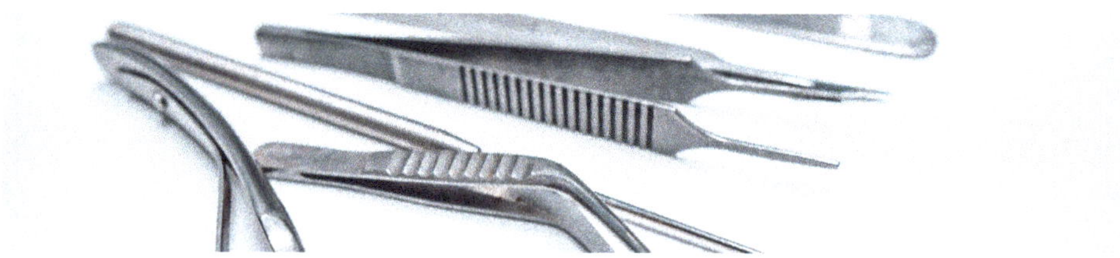

Scope of Certificate: The scope of the ISO certificate depends on how your Organization would like to have keeping in view of future. Some want to be specific by listing the products; some to be general covering the broad category of products. Here are some examples:

1. Design, development assemble, manufacture, distribute and service of life science instruments

2. Design, assemble/manufacture, sales & service of medical instruments used in cardiology including ECG Recorders, Bedside monitors

3. Design, development, manufacture, distribute and repair of medical device instruments intended to use in genetic analysis such as DNA Amplifiers, Genetic Analyzers, etc.

Distributer

Natural or legal person in the supply chain who, on his own behalf, furthers the availability of a medical device to the end user

Importer

natural or legal person in the supply chain who is the first in a supply chain to make a medical device,
manufactured in another country or jurisdiction, available in the country or jurisdiction where it is to
be marketed

Manufacturer

natural or legal person with responsibility for design and/or manufacture of a medical device with the
intention of making the medical device available for use, under his name; whether or not such a medical
device is designed and/or manufactured by that person himself or on his behalf by another person(s)

Device The term "device" is used throughout the MDSAP processes. For the purpose of applying the MDSAP processes, and to accommodate nuances in the regulatory systems of the participating Regulatory Authorities, the use of the term "device" is to refer to any product that is capable of functioning as a medical device, whether or not it is packaged, labeled, or sterilized.

Supplier: Organization or individual that enters into an agreement with the acquirer or integrator for the supply of a product or service.

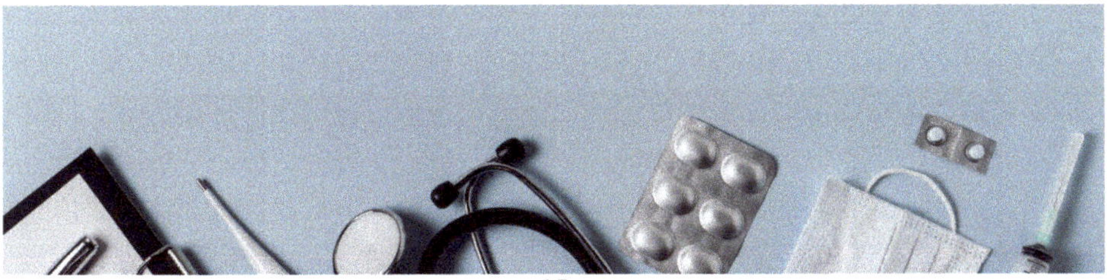

We can categorized the suppliers to:

- Critical Supplier
- Non-critical supplier

A purchased or otherwise obtained "product" or "service" is an outsourced product or service. In addition, a "supplier" is anyone that is independent from the medical device organization's quality management system. This includes a supplier that may be part of the same corporation as the medical device organization but operates under a separate quality management system from the audited medical device organization.

Critical Suppliers:

For the purposes of MDSAP, "critical suppliers" include, but are not limited to;

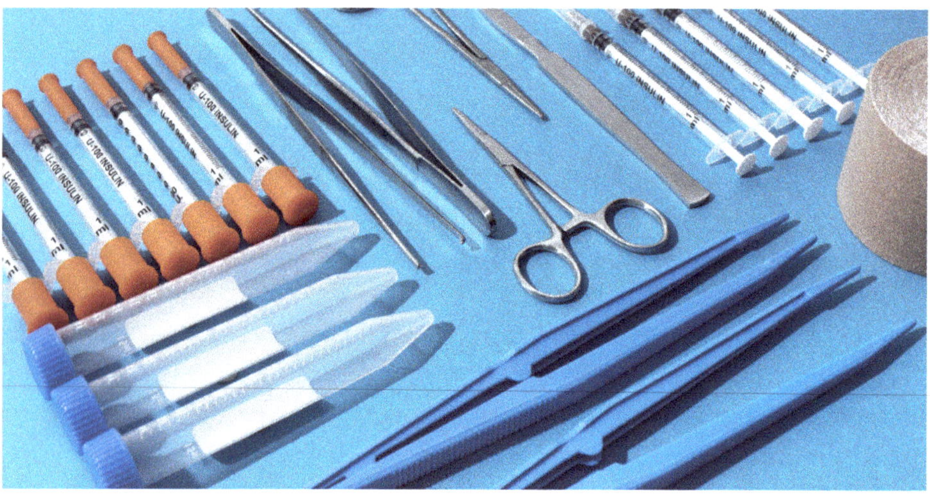

- Those entities that supply the organization with finished devices, i.e. a device, or accessory to any device, that is suitable for use or capable of functioning, whether or not it is packaged, labeled, or sterilized;
- Suppliers of products, including services, that impact design outputs that are essential for the proper functioning of the device; and
- Suppliers of products and services that require process validation.

Not all MDSAP participating regulatory authorities require, or make use of, certification documents that relate to a medical device organization's QMS. The terms "certification" and "recertification" appear within this volumes to maintain consistency with the terminology used within ISO/IEC 17021-1:2015 Conformity assessment – Requirements for bodies providing audit and certification of management systems.

MDSAP Audit

The Medical Device Single Audit Program is based on a three (3) year audit cycle. The Initial Audit, also referred to as the "Initial Certification Audit" is a complete audit of a medical device organization's quality management system (QMS) consisting of:

Stage 1 Audit (17021-1:2015 – Cl 9.3.1.2)

Stage 2 Audit (17021-1:2015 – Cl 9.3.1.3)

The initial Audit is followed by a partial Surveillance Audit (17021-1:2015 – Cl 9.6.2.2) in each of the following two (2) years and a complete Re-audit, also referred to as a "Recertification Audit" (17021-1:2015 – Cl 9.6.3.2) in the third (3rd) year. A recertification audit may also include a Stage 1 audit if there have been significant changes to the QMS that have not been otherwise adequately assessed.

Initial Certification Audit

Initial Certification audit consists of: Stage 1 & 2

Stage 1

Documentation review, evaluation of preparedness for Stage 2 audit, etc.

A Stage 1 audit shall be conducted in accordance with Clause 9.3.1.2 of ISO/IEC 17021-1:2015 and all applicable MDSAP Audit Process tasks and regulatory requirements.

From an MDSAP perspective, the primary purposes of a Stage 1 audit are (1) to determine if QMS documentation required by ISO 13485:2016 - Clauses 4.2.1 and other applicable MDSAP documentation requirements have been adequately defined, and documented; (2) to assess the medical device organization's preparedness for a Stage 2 audit; (3) to provide a focus for planning a Stage 2 audit; and, (4) to collect information regarding the scope of the quality management system and other aspects of the medical device organization.

Portions of a Stage 1 audit (e.g. documentation review) may be performed at a site other than the site(s) of the medical device organization seeking initial certification.

The outcome of the Stage 1 audit will assist the MDSAP recognized Auditing Organization in its determination of the readiness of the medical device organization to undergo a Stage 2 audit. The Au-

diting Organization shall determine how best to accomplish tasks of Stage 1 and Stage 2 with regards to off-site documentation and record review and on-site verifications. Hence portions of a Stage 1 audit (e.g. documentation review) may be performed at a site other than the site(s) of the medical device organization seeking initial certification. In practice it is intended that the Auditing Organization may combine elements of Stage 1 and Stage 2 to allow for a single on-site visit for the initial audit or re-audit of the medical device organization.

Stage 2

Evaluation of QMS Implementation and Effectiveness

A Stage 2 audit shall be conducted in accordance with Clause 9.3.1.3 of ISO/IEC 17021-1:2015 and using all applicable MDSAP Audit Process tasks.

The purpose of a Stage 2 audit is to determine if all applicable requirements of ISO 13485:2016 and the relevant regulatory requirements from participating regulatory authorities have been implemented. Stage 2 audit objectives shall specifically include an evaluation of:

- The effectiveness of the medical device organization's QMS incorporating the applicable regulatory requirements;
- Product/process related technologies (e.g. injection molding, sterilization);
- Adequate product technical documentation in relation to relevant regulatory requirements; and,
- The medical device organization's ability to comply with these requirements.

Surveillance Audits

(1st and 2nd Surveillance Audits):

A Surveillance Audit shall be conducted in accordance with Clause 9.6.2.2 of ISO/IEC 17021-1:2015 and clause 9.6.2 of IMDRF/MDSAP WG/N3:2016 and using applicable MDSAP Audit Process tasks. The purpose of a series of surveillance audits is to assure that all applicable requirements of ISO 13485:2016 and the relevant regulatory requirements from participating regulatory authorities are audited during the cycle of a three year audit program for the medical device organization. Surveillance audit objectives during the audit cycle shall specifically include evaluation of:

- The effectiveness of the medical device organization's QMS incorporating the applicable regulatory requirements.
- The medical device organization's ability to comply with these requirements; and
- New or changed product/process related technologies; and,
- New or amended product technical documentation in relation to relevant regulatory requirements.

In addition, surveillance audits shall include a review of issues related to medical device safety and effectiveness since the last audit such as complaints, problem reports, vigilance reports, and recalls/field corrections/advisory notices.

Re-audit (Recertification Audits)

A Re-audit (Recertification Audit) shall be conducted in accordance with Clause 9.6.3 of ISO/IEC 17021-1:2015 and using all applicable MDSAP Audit Process tasks. The purpose of a re-audit is

SILOSA
SCG
All Rights Reserved by:
Silosa Consulting Group Inc.
Sample Q
as a P
ISO 13
MD
Your Company
Quality Plan
Your Company
Organizational Chart
Your
Product
Vision
Mis

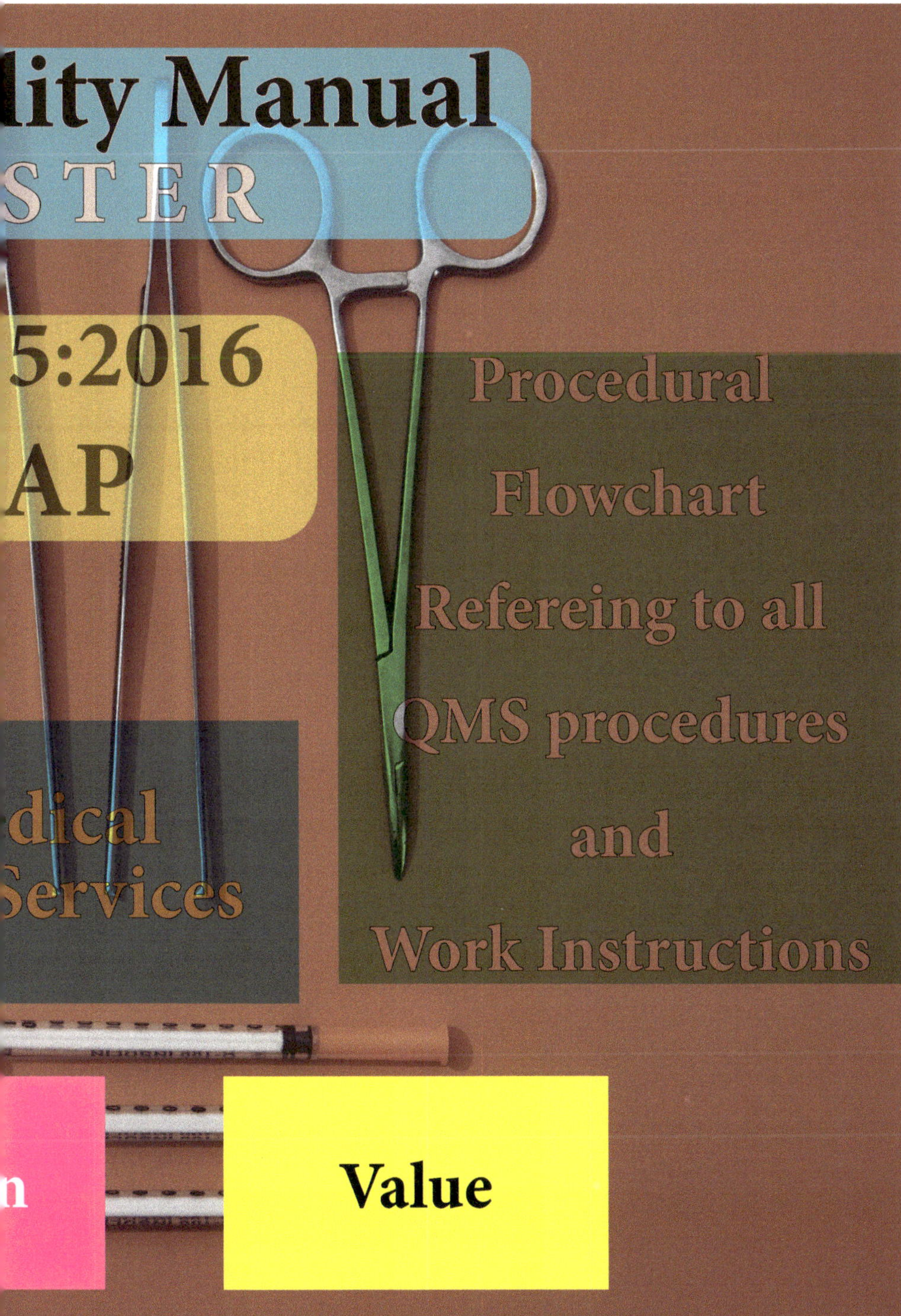
lity Manual
STER
5:2016
AP
Procedural
Flowchart
Refereing to all
QMS procedures
and
Work Instructions
dical
Services
n
Value

to confirm the continued relevance, applicability and suitability of the medical device organization's QMS (as a whole), to satisfy all applicable requirements of ISO 13485:2016 and the relevant regulatory requirements from participating regulatory authorities, with respect to the scope of certification. Recertification audit objectives shall specifically include evaluation of:
- the effectiveness of the medical device organization's QMS incorporating the applicable regulatory requirements
- product/process related technologies (e.g. injection molding, sterilization)
- adequate product technical documentation in relation to relevant regulatory requirements
- the medical device organization's continued fulfillment of these requirements.

Audits Conducted by Regulatory Authorities

Audits may also be conducted by MDSAP participating regulatory authorities at any time and for a range of reasons including (1) "For Cause" due to information obtained by the regulatory authority, (2) as follow up to the findings of a previous audit, and (3) to confirm the effective implementation of MDSAP requirements by MDSAP recognized auditing organizations.

The purpose of audits conducted by regulatory authorities is to ensure appropriate oversight of a recognized MDSAP Auditing Organization's audit activities, as an alternative means of assessing medical device organizations that have been identified as undertaking high risk manufacturing processes and have not been adequately audited, where sufficient detail regarding audited processes has not been included in an audit report, or where there is a history of low compliance with QMS or regulatory requirements.

MDSAP
ISO 13485 Requirements &

Management

The intent of the Management Process is to provide adequate resources for device design, manufacturing, quality assurance, distribution, installation, and servicing activities; to assure the quality management system is functioning properly and effectively; and to monitor the quality management system and make necessary adjustments. A quality management system that has been implemented effectively and is monitored to identify and address existing and potential problems is more likely to produce medical devices that function as intended.

The management representative is responsible for ensuring that the requirements of the quality management system have been effectively defined, documented, implemented, and maintained. Prior to the audit of a process, it may be helpful to interview the management representative (or designee) to obtain an overview of the process and a feel for management's knowledge and understanding of the process.

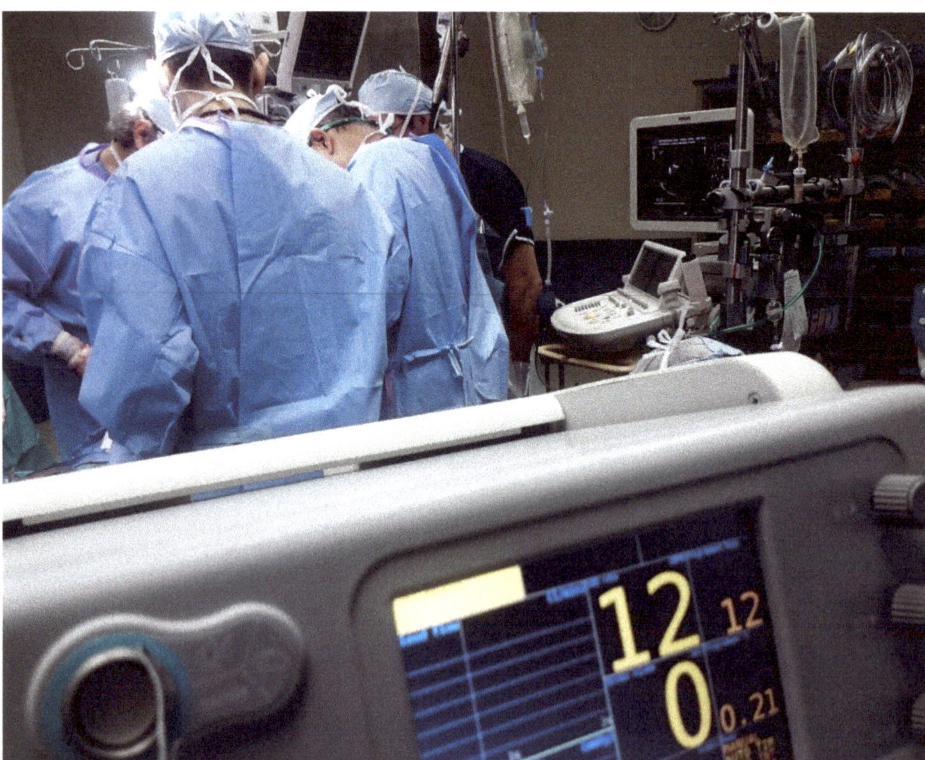

Clause and Regulation:
ISO 13485:2016: 4.1.1, 4.1.2, 4.1.3, 4.2.2, 4.1.4, 5.4.2;
Health Canada Medical Devices Regulations
(SOR /98-282)
FDA: 21 CFR 820.20

Verify that the roles and responsibilities of any regulatory correspondents, importers, distributors, or providers of a service are clearly documented in the medical device organization's quality management system and are qualified as suppliers and controlled, as appropriate.

During the audit of the other MDSAP processes, the audit team will have the opportunity to assess whether management is appropriately carrying out its responsibilities; whether the quality policy is understood, implemented, and maintained at all levels of the medical device organization; if the necessary resources are being provided to maintain an effective quality management system; if the management representative has the necessary responsibilities and authorities; the adequacy of the organizational structure; and whether management reviews and quality audits are effective, etc.

Remember that a quality management system that has been implemented effectively, monitored to identify and address existing and potential problems, and has an integrated risk management process utilizing risk-based decision-making is more likely to produce medical devices that function as intended.

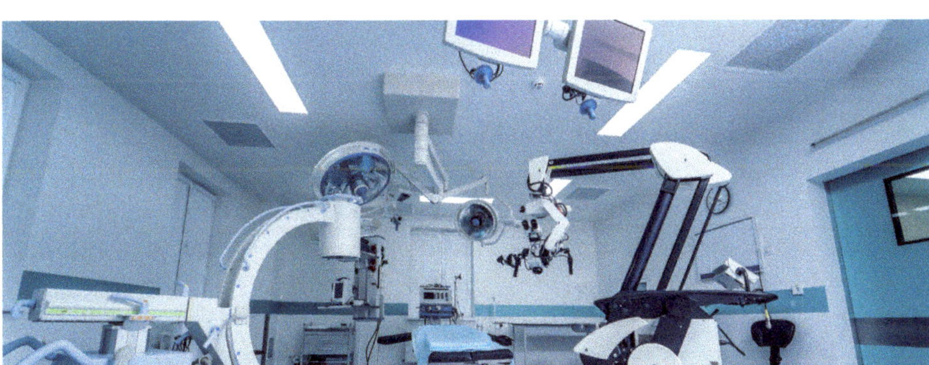

Device Marketing Authorization and Facility Registration

The Device Marketing Authorization and Facility Registration process may be audited as a linkage from the Management process and/ or the Design and Development process.

ISO: ISO 13485:2016: 4.1.1, 4.2.1, 5.2, 7.2.1, 7.2.3
Health Canada Medical Devices Regulations
(SOR /98-282)

Manufacturer means a person who sells a medical device under their own name, or under a trade-mark, design, trade name or other name or mark owned or controlled by the person, and who is responsible for designing, manufacturing, assembling, processing, labeling, packaging, refurbishing or modifying the device, or for assigning to it a purpose, whether those tasks are performed by that person or on their behalf [CMDR 1].

No person shall import or sell a Class II, III or IV medical device unless the manufacturer of the device holds a license in respect of that device or, if the medical device has been subjected to a change described in section 34, an amended medical device license [CMDR 26]. An application for a medical device license shall be submitted to the Minister by the manufacturer of the medical device in a format established by the Minister [CMDR 32].

An application for a medical device license shall include a copy of a quality management system certificate certifying that the quality management system under which the medical device is manufactured (class II) or designed and manufacturer (class III or IV) satisfies National Standard of Canada CAN/CSA-ISO 13485:2016. [CMDR 32(2)(f); 32(3)(j); 32(4)(p)].

No person shall import or sell a Class II, III or IV medical device unless the Manufacturer of the device holds a license in respect of that device or, if the medical device has been subjected to a change described in section 34 - an amended medical device license [CMDR 26].

If the Manufacturer proposes to make one or more changes, the Manufacturer shall submit to the Minister, in a format established by the Minister, an application for a medical device license amendment including the information and documents set out in section 32 that are relevant to the change
[CMDR 34].
Every Manufacturer of a licensed medical device shall, annually before November 1 and in a form authorized by the Minister, furnish the Minister with a statement signed by the Manufacturer or by a person authorized to sign on the Manufacturer's behalf describing any change to the information and documents supplied by the Manufacturer with respect to the device, other than those to be submitted under section 34 or 43.1
[CMDR 43].
If the holder of a medical device license discontinues the sale of the medical device in Canada, the licensee shall inform the Minister within 30 days after the discontinuance, and the license shall be cancelled at the time that the Minister is informed
[CMDR 43(3)].
Subject to section 34, if a new or modified quality management system certificate is issued in respect of a licensed medical device, the Manufacturer of the device shall submit a copy of the certificate to the Minister within 30 days after it is issued
[CMDR 43.1].

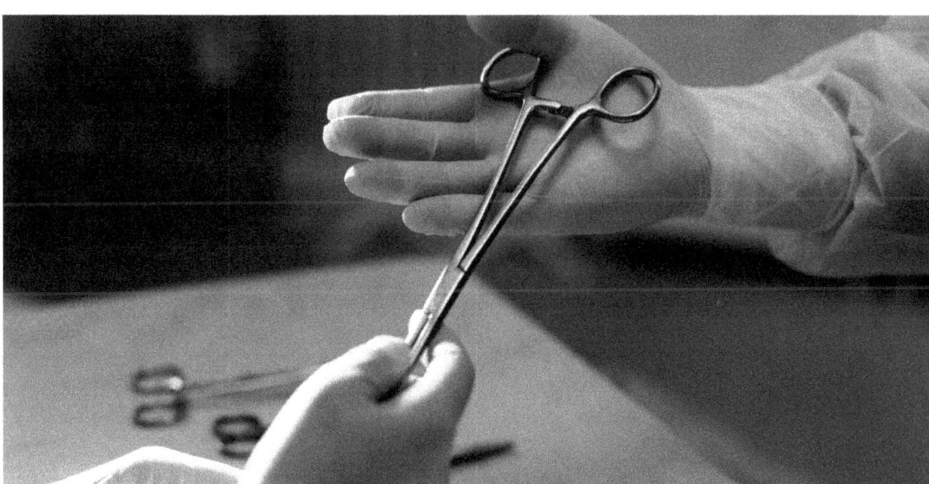

Measurement, Analysis and Improvement

One of the most important activities in the quality management system is the identification of existing and potential causes of product and quality problems. Such causes must be identified so that appropriate and effective corrective or preventive actions can take place. These activities are carried out under the Measurement, Analysis and Improvement process.

Clause and Regulation
ISO: ISO 13485:2016: 4.2.1, 8.1, 8.2.1, 8.2.6, 8.5
HC: CMDR 57-58
HC: CMDR 59-61.1
HC: CMDR 63-65.1

Verify that the Manufacturer has a process or procedure for identifying a "significant change" to a class III or IV device. Verify that information about "significant changes" is submitted in a medical device license amendment application
[CMDR 1, 34].

Verify that the Manufacturer maintains records of reported problems related to the performance characteristics or safety of a device, including any consumer complaints received by the Manufacturer after the device was first sold in Canada, and all actions taken by the Manufacturer in response to the problems referred to in the complaints .
[CMDR Section 57].

Verify that the Manufacturer has established and implemented documented procedures that will enable it to carry out an effective and timely investigation of the problem reports through the customer complaints, and to carry out an effective and timely recall of the device
[CMDR Section 58].

Medical Device Adverse Events and Advisory Notices Reporting
The Medical Device Adverse Events and Advisory Notices Reporting process may be audited as a linkage from the Measurement, Analysis and Improvement process.

Clause and Regulation
ISO: ISO 13485:2016: 4.2.1, 7.2.3, 8.2.2, 8.2.3, 8.3.3
SOR/98-282, Section 59-61.1
SOR/98-282, Section 63 – 65.1

Medical Device Regulations SOR/98-282, Section 59-61.1
Verify that the Manufacturer and the importer of a medical device make a preliminary and final report to the minister concerning any incident occurring inside or outside Canada involving a device sold in Canada:
- Related to the failure of the device or deterioration in its effectiveness or any inadequacy in its labeling or in its directions for use.
- Has led to death or serious deterioration in the state of health of a patient, user, or other person, or could do so if it were to occur [CMDR 59].
Verify that the Manufacturer or other person becoming aware of an event that led to the death or serious deterioration in the state of health of a patient, a user, or other person provides information in a preliminary report within 10 days after the person becomes aware of the event or occurrence [CMDR 60 (1) (a) (i)].
Verify that the Manufacturer or other person becoming aware of an event that the recurrence of which might lead to the death or serious deterioration in the state of health of a patient, a user, or other person provides information in a preliminary report within 30 days after the person becomes aware of the event or occurrence [CMDR 60 (1) (a) (ii)].
Note: The requirement to report incidents meeting the requirements of section 59.(1) that occur outside of Canada does not apply unless the Manufacturer has indicated, to a regulatory agency of the country in which the incident occurred, the Manufacturer's intention to take corrective action, or unless the regulatory agency has required the Manufacturer to take corrective action. [CMDR 59.(2)]
Verify that Manufacturer has made effective arrangements to sub-

mit preliminary reports to the Minister and that the reports contain [CMDR 60 (2)]:
- the identifier of any medical device that is part of a system, test kit, medical device group, medical device family or medical device group family
- if the report is made by:
- the Manufacturer: the name and address of that Manufacturer and of any known importer, and the name, title and telephone and facsimile numbers of a representative of the Manufacturer to contact for any information concerning the incident, or
- the importer of the device: the name and address of the importer and of the Manufacturer, and the name, title and telephone and facsimile numbers of a representative of the importer to contact for any information concerning the incident.
- the date on which the incident came to the attention of the Manufacturer or importer
the details known in respect of the incident, including the date on which the incident occurred and the consequences for the patient, user or other person
- the name, address and telephone number, if known, of the person who reported the incident to the Manufacturer or importer
- the identity of any other medical devices or accessories involved in the incident, if known
- the Manufacturer's or importer's preliminary comments with respect to the incident
- the course of action, including an investigation, that the Manufacturer or importer proposes to follow in respect of the incident and a timetable for carrying out any proposed action and for submitting a final report
- a statement indicating whether a previous report has been made to the Minister with respect to the device and, if so, the date of the report.

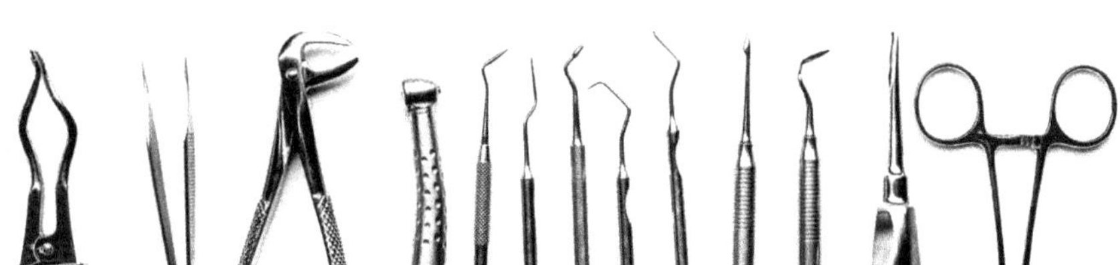

If a preliminary report required by section 60 is submitted to the Minister and/or Importer, verify that the Manufacturer has submitted a final report to the Minister in writing in accordance with the timetable established under CMDR 60(2)(h) and the final report contains
[CMDR 61(1)]:
- a description of the incident, including the number of persons who have experienced a serious deterioration in the state of their health or who have died
- a detailed explanation of the cause of the incident and a justification for the actions taken in respect of the incident
- any actions taken as a result of the investigation, which may include:
- increased post-market surveillance of the device
- corrective and preventive action respecting the design and manufacture of the device, and recall of the device.

If the reports required by section 60 and 61 are submitted to the Minister just by the Importer, verify that the Manufacturer has advised the Minister in writing that the reports the Manufacturer and importer would have submitted were identical and that the Manufacturer has permitted the importer to prepare and submit reports to the Minister on the Manufacturer's behalf [CMDR 61.1].

Verify that the Manufacturer and the importer of a medical device, on or before undertaking a recall of a device provide the minister with the following information [CMDR 64]:
- the name of the device and its identifier, including the identifier of any medical device that is part of a system, test kit, medical device group, medical device family or medical device group family
- the name and address of the Manufacturer and importer, and the name and address of the establishment where the device was manufactured, if different from that of the Manufacturer
- the reason for the recall, the nature of the defectiveness or possible defectiveness and the date on and circumstances under which the defectiveness or possible defectiveness was discovered
- an evaluation of the risk associated with the defectiveness or possible defectiveness
- the number of affected units of the device that the Manufacturer or importer:
- manufactured in Canada,

- imported into Canada,
- sold in Canada.
- the period during which the affected units of the device were distributed in Canada by the Manufacturer or importer
- the name of each person to whom the affected device was sold by the Manufacturer or importer and the number of units of the device sold to each person
- a copy of any communication issued with respect to the recall
- the proposed strategy for conducting the recall, including the date for beginning the recall, information as to how and when the Minister will be informed of the progress of the recall and the proposed date for its completion
- the proposed action to prevent a recurrence of the problem
- the name, title and telephone number of the representative of the Manufacturer or importer to contact for any information concerning the recall.

Verify that as soon as possible after the completion of the recall the Manufacturer and the importer reports to the minister the results of the recall and the action taken to prevent a recurrence of the problem [CMDR 65]. If the reports required by section 64 and 65 are submitted to the Minister just by the Importer, verify that the Manufacturer has advised the Minister in writing that the reports the Manufacturer and importer would have submitted were identical and that the Manufacturer has permitted the importer to prepare and submit reports to the Minister on the Manufacturer's behalf [CMDR 65.1].

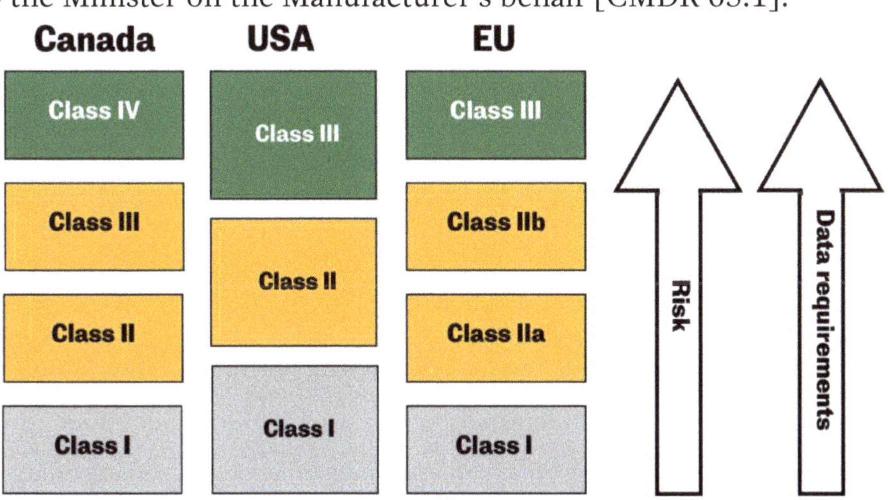

Design and Development

The purpose of the Design and Development process is to control the design of a medical device and to assure that the device meets user needs, intended use, and its specified requirements. Attention to design and development planning, identifying design inputs, developing design outputs, verifying that design outputs meet design inputs, validating the design, controlling design changes, reviewing design results, transferring the design to production, and compiling the appropriate records will help a medical device organization assure that resulting designs will meet user needs, intended uses, and requirements. Review of the Design and Development process will also provide an opportunity to evaluate how the medical device organization has utilized risk management activities to ensure design inputs are comprehensive and meet user needs, to confirm that risk control measures that were planned have been implemented in the design, and to verify that risk control measures are effective in controlling or reducing risk.

Clause and Regulation
ISO: ISO 13485:2016: 4.1.1, 4.2.1, 7.1, 7.3.10, 7.3.2
HC CMDR 9, 10 to 20, 32, 21-23, 66-68

With respect to Class II devices that are not subject to Design and Development controls, verify that the manufacturer has objective evidence to establish that Class II devices meet the safety and effectiveness requirements of section 10 to 20
[CMDR 9, 10 to 20].

Verify that Manufacturers of Class IV devices maintain a quality plan that sets out the specific quality practices, resources, and sequence of activities relevant to the device
[CMDR 32].

Production and Service Controls

The purpose of the Production and Service Controls process is to manufacture products that meet specifications. Developing processes that are adequate to produce devices that meet specifications, validating (or fully verifying the results of) those processes, and monitoring and controlling those processes are all steps that help assure the result will be devices that meet specified requirements. After completing the audit of the medical device organization's Production and Service Controls process, the audit team will return to the Management process to make a final decision of whether top management ensures that an adequate and effective quality management system has been established and maintained at the medical device organization.

Clause and Regulation
ISO: ISO 13485:2016: 7.1, 7.2.1, 7.5.1, 8.2.5, 8.2.6
HC: CMDR 9(2), 14, 17, 21-23, 52-56, 66-68

Verify that the Manufacturer maintains objective evidence that devices meet the safety and effectiveness requirements. [CMDR 9(2)].
Verify that devices sold in Canada have labeling that conforms to Canadian English and French language requirements and contains the Manufacturer's name and address, device identifier, control number (for Class III and IV devices), contents of packaging, sterility, expiry, intended use, directions for use and any special storage conditions [CMDR 21-23].
Verify that the Manufacturer maintains distribution records in respect of a device that will permit a complete and rapid withdrawal of the device from the market [CMDR 52-56].
Verify that the Manufacturer has identified Schedule 2 implants and provides implant registration cards with devices or employs another suitable system approved by Health Canada [CMDR 66-68].
Verify that the Manufacturer of devices that are listed on Schedule 2 of the Medical Devices Regulations maintains distribution records of these devices as well as any information received on implant registration cards related to these Schedule 2 devices [CMDR 54].

Reviewing a validation

During review of a validation study, determine when applicable whether:

- The instruments used to generate the data were properly calibrated and maintained
- Predetermined product and process specifications were established
- Sampling plans used to collect test samples are based on a statistically valid rationale
- Data demonstrates predetermined specifications were met consistently
- Process tolerance limits were challenged
- Process equipment was properly installed, adjusted, and maintained
- Process monitoring instruments were properly calibrated and maintained
- Changes to the validated process were appropriately challenged (if applicable)
- Process operators were appropriately qualified.

Purchasing

The intent of the Purchasing process is to ensure that purchased, subcontracted, or otherwise received products and services conform to specified requirements. The medical device organization is expected to establish and maintain documented controls for planning and performing purchasing activities. The controls necessary depend on the effect of the product on the quality, safety, and effectiveness of the finished device. Effective purchasing processes incorporate purchasing requirements and specifications, the selection of acceptable suppliers based on the capability of the suppliers to provide acceptable product, the performance of necessary acceptance activities, and maintenance of the required quality records.

The management representative is responsible for ensuring that the requirements of the quality management system have been effectively defined, documented, implemented, and maintained. Prior to the audit of a process, it may be helpful to interview the management representative to obtain an overview of the process and a feel for management's knowledge and understanding of the process.

Clause and Regulation
ISO: ISO: ISO 13485:2016: 4.1.2, 4.1.3, 4.1.5, 7.1, 7.4.1, 7.4.2, 7.4.3

Planning

In planning product realization, the medical device organization must determine as appropriate the quality objectives and requirements for the purchased products, the processes, documents, and resources specific to the purchased products, the criteria for purchased product acceptance, and the required verification, monitoring, inspection, and test activities specific to the purchased products. Planning of product realization often begins in the design and development of the product, including the translation of the design into production specifications. The translation of the design into production specifications includes the establishment of specified requirements for purchased product.

Quality objectives

Quality objectives are typically expressed as a measurable target or goal. The planning of product realization should include consideration of how the purchased product, the criteria for purchased product acceptance, and the required verification, monitoring, inspection, and test activities specific to the purchased product will achieve the quality objectives.

- Some examples of QOB include:
- Number of complaints -v- number of parts shipped
- On-time delivery %
- Supplier parts rejected
- Comparison of internal audit findings -v- external audit findings
- Achieving a certain accuracy if you're developing product - based software
- Annual Post-market surveillance
- Auditing our system for regulatory compliance

Some managers believe that the reward for hard work should be a paycheck. That's sort of like telling your children that they get to eat for doing something you're proud of. Employees are not children, but you are responsible for developing them into more valuable employees so that they can be promoted. If there is no incentive, your team will not be engaged. Therefore, pick a reward that is proportional to the bottom-line impact. Five percent of the bottom-line impact is what I like to target, but you would be amazed at how effective a few small rewards at each milestone can be. If you have trouble getting management approval for rewards, remind your boss of the bottom-line impact and link the rewards closely to the impact.

MDSAP
Audit Checklist

	REQUIREMENTS
4 Quality management system	
4.1 General requirements	
	Has the organization:
	a) identified the processes needed for the quality management system and their application throughout the organization (see 1.2)?
	b) determined the sequence and interaction of these processes?
	c) determined criteria and methods needed to ensure that both the operation and control of these processes are effective?
	d) ensured the availability of resources and information necessary to support the operation and monitoring of their processes?
	e) monitored, measured, and analyzed these processes
	f) implemented actions necessary to achieve planned results and maintain the effectiveness of these processes?
	Does the organization manage these processes in accordance with the requirements of this International Standard?
	Where an organization chooses to outsource any process that affects product conformity with requirements, does the organization ensure control over such processes?
	Is the control of such outsourced processes identified within the quality management system? (see 8.5.1)
Canada	Verify that the roles and responsibilities of any regulatory correspondents, importers, distributors, or providers of a service are clearly documented in the organization's quality management system and are qualified as suppliers and controlled.
4.2.1 General	

Doc. Reference	Adequate?	Stage I (clauses marked *)	Stage II
	Y/N	Initial›s	Initial›s

	REQUIREMENTS	
	Does the quality management system documentation include:	
	a) documented statements of a quality policy and quality objectives?	
	b) a quality manual?	
	c) documented procedures required by this international standard?	
	d) documents needed by the organization to ensure the effective planning, operation and control of its processes?	
	e) records required by this International Standard (see 4.2.4)?	
	f) any other documentation specified by national or regional regulations?	
	Has the organization established and maintained a file for each type or model of medical device either containing or identifying documents defining product specifications and quality system requirements (see 4.2.3)?	
	Do these documents define the complete manufacturing process and, if applicable, installation and servicing?	
Canada	With respect to Class II devices that are not subject to Design and Development controls, verify that the manufacturer has objective evidence to establish that Class II devices meet the safety and effectiveness requirements of section 10 to 20 [CMDR 10 ,9 to 20].	

Doc. Reference	Adequate?	Stage I (clauses marked *)	Stage II

	REQUIREMENTS	
Canada	Verify that the manufacturer maintains distribution records that contain sufficient information to permit complete and rapid withdrawal of the medical device from the market [CMDR 53-52].	
	Verify that distribution records of a device are retained by the manufacturer in a manner that will allow for timely retrieval, for the longer of (a) the projected useful life of the device; and (b) two years after the date the device was shipped [CMDR 56-55].	
EU	Does the file contain or refer to the location of objective evidence establishing the safety and effectiveness of the device as required by Annex 1 of the MDD? (MDD Annex I)	
4.2.2 Quality manual		
	Has the organization established and maintained a quality manual that includes:	
	a) the scope of the quality management system, including details of and justification for any exclusion and/or non-application (see 1.2)?	
	b) the documented procedures established for the quality management system, or reference to them?	
	c) a description of the interaction between the processes of the quality management system?	
	Does the quality manual outline the structure of the documentation used in the quality management system?	
Canada	Verify that the manufacturer has defined, documented, and implemented processes to ensure that devices are licensed prior to sale [CMDR Sections 43 ,34 ,32 ,26	
	Verify that the manufacturer has defined, documented and implemented processes to ensure that any new or modified quality management system certificate issued to the manufacturer for regulatory purposes is submitted to the Minister within 30 days after it is issued [CMDR Section 43.1].	

Doc. Reference	Adequate?	Stage I (clauses marked *)	Stage II

	REQUIREMENTS	
EU	Are the applicable sections of the Medical Device Directive (MDD) included in the specified requirements throughout the documented quality system? Interpretation: A statement only indicating compliance/conformity with the relevant international or EU regulatory requirements is not acceptable.	
4.2.3 Control of documents		
	Are documents required by the quality management system controlled?	
	Is a documented procedure established to define the controls needed:	
	a) To review and approve documents for adequacy prior to issue?	
	b) To review and update as necessary and re-approve documents?	
	c) To ensure that changes and the current revision status of documents are identified?	
	d) To ensure that relevant versions of applicable documents are available at points of use?	
	e) To ensure the documents remain legible and readily identifiable?	
	f) To ensure that documents of external origin are identified and their distribution controlled?	
	g) To prevent the unintended use of obsolete documents and to apply suitable identification to them if they are retained for any purpose?	
	Does the organization ensure that changes to documents are reviewed and approved either by the original approving function or another designated function which has access to pertinent background information upon which to base its decisions?	

Doc. Reference	Adequate?	Stage I (clauses marked *)	Stage II

	REQUIREMENTS
	Does the organization define the period for which at least one copy of obsolete controlled documents shall be retained?
	Does this period ensure that documents to which medical devices have been manufactured and tested are available for at least the lifetime of the medical device as defined by the organization, but not less than the retention period of any resulting record (see 4.2.4), or as specified by relevant regulatory requirements?
4.2.4 Control of quality records	
	Are records established and maintained to provide evidence of conformity to requirements and of the effective operation of the quality management system?
	Do records remain legible, readily identifiable and retrievable?
	Has a documented procedure been established to define the controls needed for the identification, storage, protection, retrieval, retention time and disposition of records?
	Does the organization retain the records for a period of time at least equivalent to the lifetime of the medical device as defined by the organization, but not less than two years from the date of product release by the organization or as specified by relevant regulatory requirements?
EU	Has the manufacturer retained for a period ending at least five years after the last product has been manufactured, the records listed in Annex II, 6.1 or Annex V, 5.1 or Annex VI, 5.1 (whichever applies)
5 Management responsibility	
5.1 Management commitment	

	Doc. Reference	Adequate?	Stage I (clauses marked *)	Stage II

	REQUIREMENTS	
	Has top management provided evidence of its commitment to the development and implementation of the quality management system and maintaining its effectiveness by:	
	a) Communicating to the organization the importance of meeting customer as well as statutory and regulatory requirements?	
	b) Establishing the quality policy?	
	c) Ensuring that quality objectives are established?	
	d) Conducting management reviews?	
	e) Ensuring the availability of resources?	
5.2 Customer focus		
	Does top management ensure that customer requirements are determined and met (see 7.2.1 and 8.2.1)?	
5.3 Quality policy		
	Does top management ensure that the quality policy	
	a) Is appropriate to the purpose of the organization?	
	b) Includes a commitment to comply with requirements and to maintain the effectiveness of the quality management system?	
	c) Provides a framework for establishing and reviewing quality objectives?	
	d) Is communicated and understood within the organization?	
	e) Is reviewed for continuing suitability?	
5.4 Planning		
5.4.1 Quality objectives		
	Does top management ensure that quality objectives, including those needed to meet requirements for product (see 7.1a), are established at relevant functions and levels within the organization?	
	Are quality objectives measurable?	

Doc. Reference	Adequate?	Stage I (clauses marked *)	Stage II

	REQUIREMENTS	
	Are quality objectives consistent with the quality policy?	
5.4.2 Quality management system planning		
	Has top management ensured that:	
	a) The planning of the quality management system is carried out in order to meet the requirements given in 4.1, as well as the quality objectives?	
	b) The integrity of the quality management system is maintained when changes to the quality management system are planned and implemented?	
5.5 Responsibility, authority and communication		
5.5.1 Responsibility and authority		
	Has top management ensured that responsibilities and authorities were defined, documented and communicated within the organization?	
	Has top management established the interrelation of all personnel who manage, perform and verify work affecting quality, and ensured the independence and authority necessary to perform these tasks?	
5.5.2 Management representative		
	Has top management appointed a member of management who, irrespective of other responsibilities, has responsibility and authority that includes:	
	a) Ensuring that processes needed for the quality management system are established, implemented, and maintained?	
	b) Reporting to top management on the performance of the quality management system and any need for improvement (see 8.5)?	
	c) Ensuring the promotion of awareness of regulatory and customer requirements throughout the organization?	
5.5.3 Internal communication		

	Doc. Reference	Adequate?	Stage I (clauses marked *)	Stage II

	REQUIREMENTS
	Has top management ensured that appropriate communication processes have been established within the organization and that communication takes place regarding the effectiveness of the quality management system?
5.6 Management review	
5.6.1 General	
	Does top management review the organization's quality management system, at planned intervals, to ensure its continuing suitability, adequacy and effectiveness?
	Does this review include assessing opportunities for improvement and the need for changes to the quality management system, including the quality policy and quality objectives?
	Are records from management reviews maintained (see 4.2.4)?
5.6.2 Review input	
	Does the input to management review include information on:
	a) Results of audits?
	b) Customer feedback?
	c) Process performance and product conformity?
	d) Status of preventive and corrective actions?
	e) Follow-up actions from previous management reviews?
	f) Changes that could affect the quality management system?
	g) Recommendations for improvement?
	h) New or revised regulatory requirements?
5.6.3 Review output	

	Doc. Reference	Adequate?	Stage I (clauses marked *)	Stage II

	REQUIREMENTS	
	Does output from the management review include any decisions and actions related to:	
	a) Improvements needed to maintain the effectiveness of the quality management system and its processes?	
	b) Improvement of product related to customer requirements?	
	c) Resource needs?	
6 Resource management		
6.1 Provision of resources		
	Does the organization determine and provide the resources needed:	
	a) To implement the quality management system and to maintain its effectiveness?	
	b) To meet regulatory and customer requirements?	
6.2 Human resources		
6.2.1 General		
	Are personnel performing work affecting product quality competent on the basis of appropriate education, training, skills and experience?	
6.2.2 Competence, awareness and training		
	Does the organization:	
	a) determine the necessary competence for personnel performing work affecting product quality?	
	b) Provide training or take other actions to satisfy these needs?	
	c) Evaluate the effectiveness of actions taken?	
	d) Ensure that its personnel are aware of the relevance and importance of their activities and how they contribute to the achievement of the quality objectives?	
	e) Maintain appropriate records of education, training, skills and experience (see 4.2.4)?	

Doc. Reference	Adequate?	Stage I (clauses marked *)	Stage II

	REQUIREMENTS	
US	Verify that resources include the assignment of trained personnel to meet the requirements of 21 CFR Part 820, including management, performance of work, assessment activities, and internal quality audits [21 CFR 820.20(b)(2)].	
6.3 Infrastructure		
	Does the organization determine, provide and maintain the infrastructure needed to achieve conformity to product requirements?	
	Infrastructure includes, as applicable:	
	a) Buildings, workspace and associated utilities	
	b) Process equipment, both hardware and software	
	c) Supporting services such as transport or communication	
	Does the organization establish documented requirements for maintenance activities, including their frequency, when such activities or lack thereof can affect product quality?	
	Are records of such maintenance maintained (see 4.2.4)?	
	Has the organization determined and does it manage the work environment needed to achieve conformity to product requirements?	
	a) Has the organization established documented requirements for health, cleanliness and clothing of personnel if contact between such personnel and the product or work environment could adversely affect the quality of the product (see 7.5.1.2.1)?	
	b) If work environment conditions can have an adverse effect on product quality, has organization established documented requirements for the work environment conditions and documented procedures or work instructions to monitor and control these work environment conditions (see 7.5.1.2.1)?	

Doc. Reference	Adequate?	Stage I (clauses marked *)	Stage II

	REQUIREMENTS	
	c) Does the organization ensure that all personnel who are required to work temporarily under special environmental conditions within the work environment are appropriately trained or supervised by a trained person [see 6.2.2 b)]?	
	d) If appropriate, are special arrangements established and documented for the control of contaminated or potentially contaminated product in order to prevent contamination of other product, the work environment or personnel (see 7.5.3.1)?	
7 Product realization		
7.1 Planning of product realization		
	Has the organization planned and developed the processes needed for product realization?	
	Is the planning of product realization consistent with the requirements of the other processes of the quality management system (see 4.1)?	
	In planning product realization, has the organization determined the following, as appropriate:	
	a) Quality objectives and requirements for products?	
	b) The need to establish processes, documents, and provide resources specific to the product?	
	c) Required verification, validation, monitoring, inspection and test activities specific to the product and the criteria for product acceptance?	
	d) Records needed to provide evidence that the realization processes and resulting product meet requirements (see 4.2.4)?	
	Is the output of this planning in a form suitable for the organization's method of operations?	

Doc. Reference	Adequate?	Stage I (clauses marked *)	Stage II

	REQUIREMENTS	
	Does the organization establish documented requirements for risk management throughout product realization and are records arising from risk management maintained (see 4.2.4)?	
Canada	Verify that the manufacturer has a process or procedure for identifying a "significant change" to a class III or IV device. Verify that information about "significant changes" is submitted in a medical device license amendment application [CMDR 34 ,1].	
EU	Does the supplier evaluate the need for risk analysis throughout the design process and maintain records of any risk analysis performed? (MDD Annex 1)	
7.2.1 Determination of requirements related to the product		
	Has the organization determined:	
	a) Requirements specified by the customer, including the requirements for delivery and post-delivery activities?	
	b) Requirements not stated by the customer but necessary for specified or intended use, where known?	
	c) Statutory and regulatory requirements related to the product?	
	d) Any additional requirements determined by the organization?	

	Doc. Reference	Adequate?	Stage I (clauses marked *)	Stage II

	REQUIREMENTS
Canada	Verify that the manufacturer has met the device licensing requirements prior to mport or sell a Class II, III or IV medical device in Canada [CMDR 32]. Manufacturer means a person who sells a medical device under their own name, or under a trade-mark, design, trade name or other name or mark owned or controlled by the person, and who is responsible for designing, manufacturing, assembling, processing, labeling, packaging, refurbishing or modifying the device, or for assigning to it a purpose, whether those tasks are performed by that person or on their behalf [CMDR 1]. No person shall import or sell a Class II, III or IV medical device unless the manufacturer of the device holds a license in respect of that device or, if the medical device has been subjected to a change described in section 34, an amended medical device license [CMDR 26]. An application for a medical device license shall be submitted to the Minister by the manufacturer of the medical device in a format established by the Minister [CMDR 32]. An application for a medical device license shall include a copy of a quality management system certificate certifying that the quality management system under which the medical device is manufactured (class II) or designed and manufacturer (class III or IV) satisfies National Standard of Canada CAN/CSA-ISO 13485:2003. [CMDR 2)32)(f); 3)32)(j); 4)32)(p)].

Doc. Reference	Adequate?	Stage I (clauses marked *)	Stage II

	REQUIREMENTS	
EU	Vefiy that manufacturing maintains files containing or refer to the location of objective evidence establishing the safety and effectiveness of the device as required by Annex 1 of the MDD. Verify that the manufacturer followed a defined and effective process to establish and maintain a file containing documents defining product specifications and quality system requirements for each newa and existing type/modes of medical devices. Since the last audit, has the manufacturer introduced new products in the EU? Has the manufacturer followed a defined and effective process to obtain an approval from the Notified Body to CE mark a product prior to selling it in the EU? (does not apply to class I devices)	
7.2.2 Review of requirements related to the product		
	Does the organization review the requirements related to the product?	
	Is this review conducted prior to the organization's commitment to supply a product to the customer (e.g. submission of tenders, acceptance of contracts or orders, acceptance of changes to contracts or orders)?	
	Does the organization ensure that:	
	a) Product requirements are defined and documented?	
	b) Contract or order requirements differing from those previously expressed are resolved?	
	c) The organization has the ability to meet the defined requirements?	
	Are records of the results of the review and actions arising from the review maintained (see 4.2.4)?	

Doc. Reference	Adequate?	Stage I (clauses marked *)	Stage II

	REQUIREMENTS	
	Where the customer provides no documented statement of requirement, are the customer requirements confirmed by the organization before acceptance?	
	Where product requirements are changed, does the organization ensure that relevant documents are amended and that relevant personnel are made aware of the changed requirements?	
7.2.3 Customer communication		
	Has the organization determined and implemented effective arrangements for communicating with customers in relation to:	
	a) Product information?	
	b) Enquiries, contracts or order handling, including amendments?	
	c) Customer feedback, including customer complaints? (see 8.2.1)	
	d) Advisory notices? (see 8.5.1)	
7.3 Design and development		
7.3.1 Design and/or development planning		
	Has the organization established documented procedures for design and development?	
	Does the organization plan and control the design and development of product?	
	During the design and development planning, does the organization determine:	
	a) The design and development stages?	
	b) The review, verification, validation and design transfer activities (see Note) that are appropriate at each design and development stage?	
	c) The responsibilities and authorities for design and development?	

Doc. Reference	Adequate?	Stage I (clauses marked *)	Stage II

	REQUIREMENTS	
	Does the organization manage the interfaces between different groups involved in design and development to ensure effective communication and clear assignment of responsibility?	
	Is planning output documented, and updated as appropriate, as the design and development progresses? (See 4.2.3)	
Canada	Verify that manufacturers of Class IV devices maintain a quality plan that sets out the specific quality practices, resources, and sequence of activities relevant to the device [CMDR 32].	
7.3.2 Design and development inputs		
	Have inputs relating to product requirements been determined and records maintained (see 4.2.4)?	
	Do these inputs include:	
	a) Functional, performance and safety requirements, according to the intended use?	
	b) Applicable statutory and regulatory requirements?	
	c) Where applicable, information derived from previous similar designs?	
	d) Other requirements essential for design and development?	
	e) Output(s) of risk management (see 7.1)?	
	Are these inputs reviewed for adequacy and approved?	
	Are requirements complete, unambiguous and not in conflict with each other?	
7.3.3 Design and development outputs		
	Are the outputs of design and development provided in a form that enables verification against the design and development input, and is it approved prior to release?	

Doc. Reference	Adequate?	Stage I (clauses marked *)	Stage II

	REQUIREMENTS
	Do design and development outputs:
	a) meet the input requirements for design and development?
	b) Provide appropriate information for purchasing, production and for service provision?
	c) Contain or reference product acceptance criteria?
	d) Specify the characteristics of the product that are essential for its safe and proper use?
	Are records of the design and development outputs maintained (see 4.2.4)?
7.3.4 Design and development review	
	At suitable stages, are systematic reviews of design and development performed in accordance with planned arrangements (see 7.3.1):
	a) To evaluate the ability of the results of design and development to meet requirements?
	b) To identify any problems and propose necessary actions?
	Do participants in such reviews include representatives of functions concerned with the design and development stage(s) being reviewed, as well as other specialist personnel (see 5.5.1 and 6.2.1)?
	Are records of the results of the reviews and any necessary actions maintained (see 4.2.4)?
EU	Does the supplier evaluate the need for risk analysis throughout the design process and maintain records of any risk analysis performed? (MDD Annex 1)
7.3.5 Design and development verification	
	Is verification performed in accordance with planned arrangements (see 7.3.1) to ensure that the design and development outputs have met the design and development input requirements?
	Are records of the results of the verification and any necessary actions maintained?

Doc. Reference	Adequate?	Stage I (clauses marked *)	Stage II

	REQUIREMENTS	
7.3.6 Design and development validation		
	Is design and development validation performed in accordance with planned arrangements (see 7.3.1) to ensure that the resulting product is capable of meeting the requirements for the specified application or intended use?	
	Is validation completed prior to the delivery or implementation of the product (see Note 1)?	
	Are records of the results of validation and any necessary actions maintained (see 4.2.4)?	
	As part of design and development validation, does the organization perform clinical evaluations and/or evaluation of performance of the medical device, as required by national or regional regulations (see Note 2)?	
7.3.7 Control of design and development changes		
	Are design and development changes identified and records maintained?	
	Are changes reviewed, verified, and validated, as appropriate, and approved before implementation?	
	Does the review of design and development changes include evaluation of the effect of the changes on constituent parts and product already delivered?	
	Are records of the results of the review of changes and any necessary actions maintained?	
Canada	Verify that the manufacturer has a process or procedure for identifying a "significant change" to a Class III or IV medical device. Verify that information about "significant changes" is submitted in a medical device licenseamendment application [CMDR 34 ,1].	
EU	Do documented procedures identify the need to report essential changes to the Notified Body, (MDD Annex II, V, VI, 3.4)	

	Doc. Reference	Adequate?	Stage I (clauses marked *)	Stage II

	REQUIREMENTS	
7.4 Purchasing		
7.4.1 Purchasing process		
	Has the organization established documented procedures to ensure that purchased product conforms to specified purchase requirements?	
	Is the type and extent of control applied to the supplier and the purchased product dependent upon the effect of the purchased product on subsequent product realization or the final product?	
	Does the organization evaluate and select suppliers based on their ability to supply product in accordance with the organization's requirements?	
	Have criteria for selection, evaluation and re-evaluation been established?	
	Are records of the results of evaluation and any necessary actions arising from the evaluation maintained (see 4.2.4)?	
Canada	Verify that any regulatory correspondent used by the manufacturer is treated as a supplier and is adequately qualified.	
7.4.2 Purchasing information		
	Does purchasing information describe the product to be purchased, including where appropriate:	
	a) Requirements for approval of product, procedures, processes and equipment?	
	b) Requirements for qualification of personnel?	
	c) Quality management system requirements?	
	Does the organization ensure the adequacy of specified purchase requirements prior to their communication to the supplier?	
	To the extent required for traceability given in 7.5.3.2, does the organization maintain relevant purchasing information, i.e. documents (see 4.2.3) and records (see 4.2.4)?	

	Doc. Reference	Adequate?	Stage I (clauses marked *)	Stage II

	REQUIREMENTS	
7.4.3 Verification of purchased product		
	Has the organization established and implemented the inspection or other activities necessary for ensuring that purchased product meets specified purchase requirements?	
	Where the organization or its customer intends to perform verification at the supplier's premises, does the organization state the intended verification arrangements and method of product release in the purchasing information?	
	Are records of the verification maintained (see 4.2.4)?	
7.5 Production and service provision		
7.5.1 Control of production and service provision		
7.5.1.1 General Requirements		
	Does the organization plan and carry out production and service provision under controlled conditions?	
	Do the controlled conditions include, as applicable:	
	a) The availability of information that describes the characteristics of the product?	
	b) The availability of documented procedures, documented requirements, work instructions, reference materials and reference measurement procedures as necessary?	
	c) The use of suitable equipment?	
	d) The availability and use of monitoring and measuring devices?	
	e) The implementation of monitoring and measurement?	
	f) The implementation of release, delivery and post-delivery activities?	
	g) The implementation of defined operations for labeling and packaging?	

Doc. Reference	Adequate?	Stage I (clauses marked *)	Stage II

	REQUIREMENTS	
	Does the organization establish and maintain a record (see 4.2.4) for each batch of medical devices that provides traceability to the extent specified in 7.5.3 and identifies the amount manufactured and amount approved for distribution?	
	Is the batch record verified and approved?	
7.5.1.2 Control of production and service provision — Specific requirements		
7.5.1.2.1 Cleanliness of product and contamination control		
	Does the organization establish documented requirements for cleanliness of product if:	
	a) Product is cleaned by the organization prior to sterilization and/or its use, or	
	b) Product is supplied non-sterile to be subjected to a cleaning process prior to sterilization and/or its use, or	
	c) Product is supplied to be used non-sterile and its cleanliness is of significance in use, or	
	d) Process agents are to be removed from product during manufacture?	
	If product is cleaned in accordance with a) or b) above, the requirements contained in 6.4 a) and 6.4 b) do not apply prior to the cleaning process.	
7.5.1.2.2 Installation activities		
	If appropriate, does the organization establish documented requirements which contain acceptance criteria for installing and verifying the installation of the medical device?	
	If the agreed customer requirements allow installation to be performed other than by the organization or its authorized agent, does the organization provide documented requirements for installation and verification?	
	Are records of installation and verification performed by the organization or its authorized agent maintained (see 4.2.4)?	

Doc. Reference	Adequate?	Stage I (clauses marked *)	Stage II

	REQUIREMENTS	
7.5.1.2.3 Servicing activities		
	If servicing is a specified requirement, does the organization establish documented procedures, work instructions and reference materials and reference measurement procedures, as necessary, for performing servicing activities and verifying that they meet the specified requirements?	
	Are records of servicing activities carried out by the organization maintained (see 4.2.4)?	
7.5.1.3 Particular requirements for sterile medical devices		
	Does the organization maintain records of the process parameters for the sterilization process which was used for each sterilization batch (see 4.2.4)?	
	Are sterilization records traceable to each production batch of medical devices (see 7.5.1.1)?	
7.5.2 Validation of processes for production and service provision		
7.5.2.1 General requirements		
	Does the organization validate any processes for production and service provision where the resulting output cannot be verified by subsequent monitoring or measurement? (this includes any processes where deficiencies become apparent only after the product is in use or the service has been delivered)	
	Does validation demonstrate the ability of the processes to achieve planned results?	
	Has the organization established arrangements for these processes including, as applicable:	
	a) Defined criteria for review and approval of the processes?	
	b) Approval of equipment and qualification of personnel?	
	c) Use of specific methods and procedures?	
	d) Requirements for records (see 4.2.4)?	
	e) Re-validation?	

Doc. Reference	Adequate?	Stage I (clauses marked *)	Stage II

	REQUIREMENTS	
	Has the organization established documented procedures for the validation of the application of computer software (and changes to such software and/or its application) for production and service provision that affect the ability of the product to conform to specified requirements?	
	Are such software applications validated prior to initial use?	
	Are records of validation maintained (see 4.2.4)?	
7.5.2.2 Particular requirements for sterile medical devices		
	Has the organization established documented procedures for the validation of sterilization processes?	
	Are sterilization processes validated prior to initial use?	
	Are records of validation of each sterilization process maintained (see 4.2.4)?	
Canada	Verify that sterilization methods for devices sold in a sterile state are validated [CMDR 17].	
7.5.3 Identification and traceability		
7.5.3.1 Identification		
	Does the organization identify the product by suitable means throughout product realization and does the organization establish documented procedures for such product identification?	
	Has the organization established documented procedures to ensure that medical devices returned to the organization are identified and distinguished from conforming product [see 6.4 d)]?	
7.5.3.2 Traceability		
7.5.3.2.1 General		
	Has the organization established documented procedures for traceability?	
	Do such procedures define the extent of product traceability and the records required (see 8.3 ,4.2.4 and 8.5)?	

	Doc. Reference	Adequate?	Stage I (clauses marked *)	Stage II

	REQUIREMENTS	
	Does the organization control and record the unique identification of the product, where traceability is a requirement (see 4.2.4)?	
Canada	Verify that the manufacturer maintains objective evidence that devices meet the safety and effectiveness requirements of the CMDR [CMDR 2)9)]. Verify that devices sold in Canada have labeling that conforms to Canadian English and French language requirements and contains the manufacturer's name and address, device identifier, control number (for Class III and IV devices), contents of packaging, sterility, expiry, intended use, directions for use and any special storage conditions [CMDR 23-21].	
EU	Does the manufacturer have procedures identifying the requirements of labeling and instructions for use as defined in MDD, Annex 1 point 13, and is regulations for CE marking included in these procedures. (MDD 3, Article 17 and Annex 12).	
	Are language requirements defined in procedures for the information identified in MDD, Annex 1point 13 for the applicable markets.	
7.5.3.2.2 Particular requirements for active implantable medical devices and i		
	In defining the records required for traceability, does the organization include records of all components, materials and work environment conditions, if these could cause the medical device not to satisfy its specified requirements?	
	Does the organization require that its agents or distributors maintain records of the distribution of medical devices to allow traceability and that such records be available for inspection?	

	Doc. Reference	Adequate?	Stage I (clauses marked *)	Stage II
ble medical devices				

	REQUIREMENTS	
	Are records of the name and address of the shipping package consignee maintained (see 4.2.4)?	
Canada	Verify that the manufacturer has identified Schedule 2 implants and provides implant registration cards with devices or employs another suitable system approved by Health Canada [CMDR 68-66]. Verify that the manufacturer of devices that are listed on Schedule 2 of the Medical Devices Regulations maintains distribution records of these devices as well as any information received on implant registration cards related to these Schedule 2 devices [CMDR 54].	
7.5.3.3 Status identification		
	Does the organization identify the product status with respect to monitoring and measurement requirements?	
	Is the identification of product status maintained throughout production, storage, installation and servicing of the product to ensure that only product that has passed the required inspections and tests (or released under an authorized concession) is dispatched, used or installed?	
7.5.4 Customer property		
	Does the organization exercise care with customer property while it is under the organization's control or being used by the organization?	
	Does the organization identify, verify, protect and safeguard customer property provided for use or incorporation into the product?	
	If any customer property is lost, damaged, or otherwise found to be unsuitable for use, is this reported to the customer and are records maintained (see 4.2.4)?	
7.5.5 Preservation of product		

Doc. Reference	Adequate?	Stage I (clauses marked *)	Stage II

	REQUIREMENTS	
	Has the organization established documented procedures or documented work instructions for preserving the conformity of product during internal processing and delivery to the intended destination?	
	Does this preservation include identification, handling, packaging, storage and protection, and also apply to the constituent parts of a product?	
	Has the organization established documented procedures or documented work instructions for the control of product with a limited shelf-life or requiring special storage conditions?	
	Are such special storage conditions controlled and recorded (see 4.2.4)?	
7.6 Control of monitoring and measuring devices		
	Does the organization determine monitoring and measuring to be undertaken and the monitoring and measuring devices needed to provide evidence of conformity of product to determined requirements (guide reference 7.2.1)?	
	Has the organization established documented procedures to ensure that monitoring and measurement can be carried out and are carried out in a manner that is consistent with the monitoring and measurement requirements?	

Doc. Reference	Adequate?	Stage I (clauses marked *)	Stage II

	REQUIREMENTS	
	Where necessary to ensure valid results, is measuring equipment:	
	a) Calibrated or verified at specified intervals, or prior to use, against measurement standards traceable to international or national measurement standards; where no such standards exist, is the basis used for calibration or verification recorded?	
	b) Adjusted or re-adjusted as necessary?	
	c) Identified to enable the calibration status to be determined?	
	d) Safeguarded from adjustments that would invalidate the measurement result?	
	e) Protected from damage and deterioration during handling, maintenance and storage?	
	Does the organization assess and record the validity of the previous measuring results when the equipment is found not to conform to requirements?	
	Does the organization take appropriate action on the equipment and any product affected?	
	Are records of the results of calibration and verification maintained (see 4.2.4)?	
	When used in the monitoring and measurement of specified requirements, is the ability of computer software to satisfy the intended application confirmed prior to initial use and reconfirmed as necessary ?	
8 Measurement, analysis and improvement		
8.1 General		

	Doc. Reference	Adequate?	Stage I (clauses marked *)	Stage II

	REQUIREMENTS	
	Does the organization plan and implement the monitoring, measurement, analysis and improvement processes needed:	
	a) To demonstrate conformity of the product?	
	b) To ensure conformity of the quality management system?	
	c) To maintain the effectiveness of the quality management system?	
	Does this include determination of applicable methods, including statistical techniques, and the extent of their use?	
8.2 Monitoring and measurement		
8.2.1 Feedback		
	As one of the measurements of the performance of the quality management system, does the organization monitor information relating to whether the organization has met customer requirements?	
	Have the methods for obtaining and using this information been determined?	
	Has the organization established a documented procedure for a feedback system [see 7.2.3 c)] to provide early warning of quality problems and for input into the corrective and preventive action processes (see 8.5.2 and 8.5.3)?	
	If national or regional regulations require the organization to gain experience from the post-production phase, does the review of this experience form part of the feedback system (see 8.5.1)?	

Doc. Reference	Adequate?	Stage I (clauses marked *)	Stage II

	REQUIREMENTS	
Canada	Verify that the manufacturer maintains records of reported problems related to the performance characteristics or safety of a device, including any consumer complaints received by the manufacturer after the device was first sold in Canada, and all actions taken by the manufacturer in response to the problems referred to in t hecomplaints [CMDR Section 57]. Verify that the manufacturer has established and implemented documented procedures that will enable it to carry out an effective and timely investigation of the problems reports through the customer complaints, andto carry out an effective and timely recall of the device [CMDR Section 58].	
8.2.2 Internal audit		
	Does the organization conduct internal audits at planned intervals to determine whether the quality management system	
	a) Conforms to the planned arrangements (see 7.1), to the requirements of this International Standard and to the quality management system requirements established by the organization?	
	b) Is effectively implemented and maintained?	
	Is an audit programme planned, taking into consideration the status and importance of the processes and areas to be audited, as well as the results of previous audits?	
	Are the audit criteria, scope, frequency and methods defined?	
	Does selection of auditors and conduct of audits ensure objectivity and impartiality of the audit process (e.g. auditors shall not audit their own work)?	
	Are the responsibilities and requirements for planning and conducting audits, and for reporting results and maintaining records (see 4.2.4) defined in a documented procedure?	

Doc. Reference	Adequate?	Stage I (clauses marked *)	Stage II

	REQUIREMENTS	
	Does management responsible for the area being audited ensure that actions are taken without undue delay to eliminate detected nonconformities and their causes?	
	Do follow-up activities include the verification of the actions taken and the reporting of verification results? (see 8.5.2)	
8.2.3 Measurement and monitoring of processes		
	Does the organization apply suitable methods for monitoring and, where applicable, measurement of the quality management system processes?	
	Do these methods demonstrate the ability of the processes to achieve planned results?	
	When planned results are not achieved, is correction and corrective action taken, as appropriate, to ensure conformity of the product?	
8.2.4 Monitoring and measurement of product		
8.2.4.1 General requirements		
	Does the organization monitor and measure the characteristics of the product to verify that product requirements have been met?	
	Is this carried out at appropriate stages of the product realization process in accordance with the planned arrangements (see 7.1) and documented procedures (see 7.5.1.1)?	
	Is evidence of conformity with the acceptance criteria maintained?	
	Do records indicate the person(s) authorizing release of the product (see 4.2.4)?	
	Does the organization ensure that product release and service delivery do not proceed until the planned arrangements (see 7.1) have been satisfactorily completed?	
8.2.4.2 Particular requirement for active implantable medical devices and imp		

	Doc. Reference	Adequate?	Stage I (clauses marked *)	Stage II
medical devices				

	REQUIREMENTS	
	Does the organization record (see 4.2.4) the identity of personnel performing any inspection or testing?	
8.3 Control of nonconforming product		
	Does the organization ensure that product which does not conform to product requirements is identified and controlled to prevent its unintended use or delivery?	
	Are the controls and related responsibilities and authorities for dealing with nonconforming product defined in a documented procedure?	
	Does the organization deal with nonconforming product by one or more of the following ways?	
	a) By taking action to eliminate the detected nonconformity	
	b) By authorizing its use, release or acceptance under concession	
	c) By taking action to preclude its original intended use or application	
	Does the organization ensure that nonconforming product is accepted by concession only if regulatory requirements are met?	
	Are records of the identity of the person(s) authorizing the concession maintained (see 4.2.4)?	
	Are records of the nature of nonconformities and any subsequent actions taken, including concessions obtained maintained (see 4.2.4)?	
	When nonconforming product is corrected, is it subject to re-verification to demonstrate conformity to the requirements?	
	When nonconforming product is detected after delivery or use has started, does the organization take action appropriate to the effects, or potential effects, of the nonconformity?	

Doc. Reference	Adequate?	Stage I (clauses marked *)	Stage II

	REQUIREMENTS	
	If product needs to be reworked (one or more times), does the organization document the rework process in a work instruction that has undergone the same authorization and approval procedure as the original work instruction?	
	Prior to authorization and approval of the work instruction, is a determination of any adverse effect of the rework upon product made and documented (see 4.2.3 and 7.5.1)?	
8.4 Analysis of data		
	Does the organization establish documented procedures to determine, collect and analyse appropriate data to demonstrate the suitability and effectiveness of the quality management system and to evaluate if improvement of the effectiveness of the quality management system can be made?	
	Does this include data generated as a result of monitoring and measurement and from other relevant sources?	
	Does the analysis of data provide information relating to:	
	a) Feedback (see 8.2.1)?	
	b) Conformity to product requirements? (See 7.2.1)	
	c) Characteristics and trends of processes and products including opportunities for preventive action?	
	d) Suppliers?	
	Are records of the results of the analysis of data maintained (see 4.2.4)?	
8.5 Improvement		
8.5.1 General		
	Does the organization identify and implement any changes necessary to ensure and maintain the continued suitability and effectiveness of the quality management system through the use of the quality policy, quality objectives, audit results, analysis of data, corrective and preventive actions and management review?	

Doc. Reference	Adequate?	Stage I (clauses marked *)	Stage II

	REQUIREMENTS	
	Does the organization establish documented procedures for the issue and implementation of advisory notices and are these procedures capable of being implemented at any time?	
Canada	Verify the manufacturer has established procedures for reporting of recalls compliant with: [CMDR 65.1 – 63 ,1 Guide to Recall of Medical Devices GUI0054-]	
EU	Are the procedures for Vigilance reporting in conformance with MDD Annex II, V, and VI (MEDDEV 1-2.12)	
	Are records of all customer complaint investigations maintained (see 4.2.4)?	
	If investigation determines that the activities outside the organization contributed to the customer complaint, is relevant information exchanged between the organizations involved (see 4.1)?	
	If any customer complaint is not followed by corrective and/or preventive action, is the reason authorized (see 5.5.1) and recorded (see 4.2.4)?	
	If national or regional regulations require notification of adverse events that meet specific reporting criteria, does the organization establish documented procedures to such notification to regulatory authorities?	
Canada	Verify that the manufacturer has established and implemented procedures compliant with requirements of: [Medical Device Regulations SOR/282-98, Section 61.1-59: Guidance Document for Mandatory Problem Reporting for Medical Devices]	
EU	Are procedures for the reporting of recall to the relevant competent authority and the NB compliant? (MDD, Annex II, V, VI, 3.1)	
8.5.2 Corrective action		

Doc. Reference	Adequate?	Stage I (clauses marked *)	Stage II

	REQUIREMENTS	
	Does the organization take action to eliminate the cause of nonconformities in order to prevent recurrence and are corrective actions appropriate to the effects of the nonconformities encountered?	
	Has a documented procedure been established to define requirements for:	
	a) Reviewing nonconformities (including customer complaints)?	
	b) Determining the causes of nonconformities?	
	c) Evaluating the need for action to ensure that nonconformities do no recur?	
	d) Determining and implementing action needed, including, if appropriate, updating documentation (see 4.2)?	
	e) Recording of the results of any investigation and of action taken (see 4.2.4)?	
	f) Reviewing the corrective action taken and its effectiveness?	
8.5.3 Preventive action		
	Does the organization determine action to eliminate the causes of potential nonconformities in order to prevent their occurrence and are preventive actions appropriate to the effects of the potential problems?	
	Has a documented procedure been established to define requirements for:	
	a) Determining potential nonconformities and their causes?	
	b) Evaluating the need for action to prevent occurrence of nonconformities?	
	c) Determining and implementing action needed?	
	d) Recording of the results of any investigations and of action taken (see 4.2.4)?	
	e) Reviewing preventive action taken and its effectiveness?	

Doc. Reference	Adequate?	Stage I (clauses marked *)	Stage II

The system is working for you (the system is fully integrated along your processes and eases your operations).

You are working for the system
(the system is beside your operations and looks as an additional burden.)

Advice for manufacturers planning certification

Firstly, you should confirm that the device(s) your company manufactures can be defined as a medical device under ISO 13485 standards; or, if you provide a medical device service, that your service is related to a product defined as a medical device. Getting ISO 13485 certification is challenging and requires commitment so, secondly, it's important that your leadership team confirms that holding the certification will add value to your company, meet its business objectives and support its strategy. While holding the full certification is not strictly necessary, as your company can still conform to and benefit from ISO 13485 standards without being externally certified, it does clearly demonstrate to all stakeholders that you comply with its requirements.

If you do want to become independently certified there are two phases; the first covers documentation, while the second implements your quality management system and audits it. You must carry out phase two within six months of completing phase one.

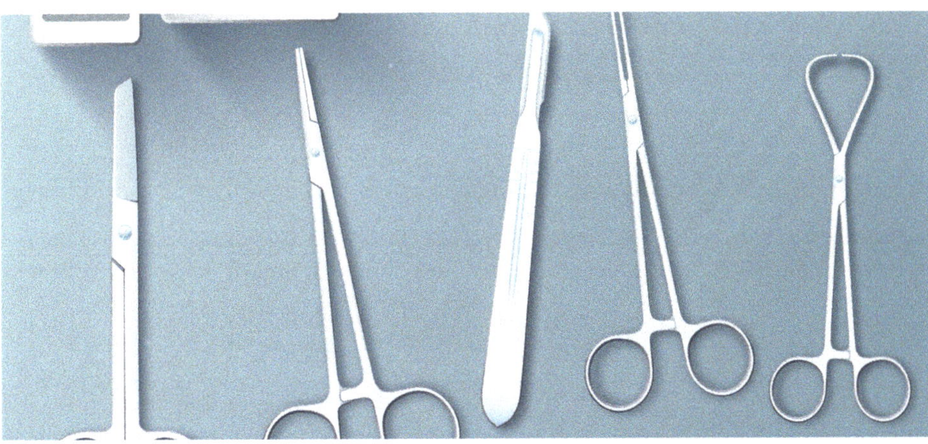

ISO 13485 certification costs

For companies who wish to be fully ISO 13485 certified, there are associated costs for external, independent certification. Although you may be under cost pressure, it's important to work with an independent auditor who will add real value to your company during the certification process. In this highly regulated industry, choose a reputable auditing company with extensive expertise rather than one that enables you to gain certification at the lowest price.

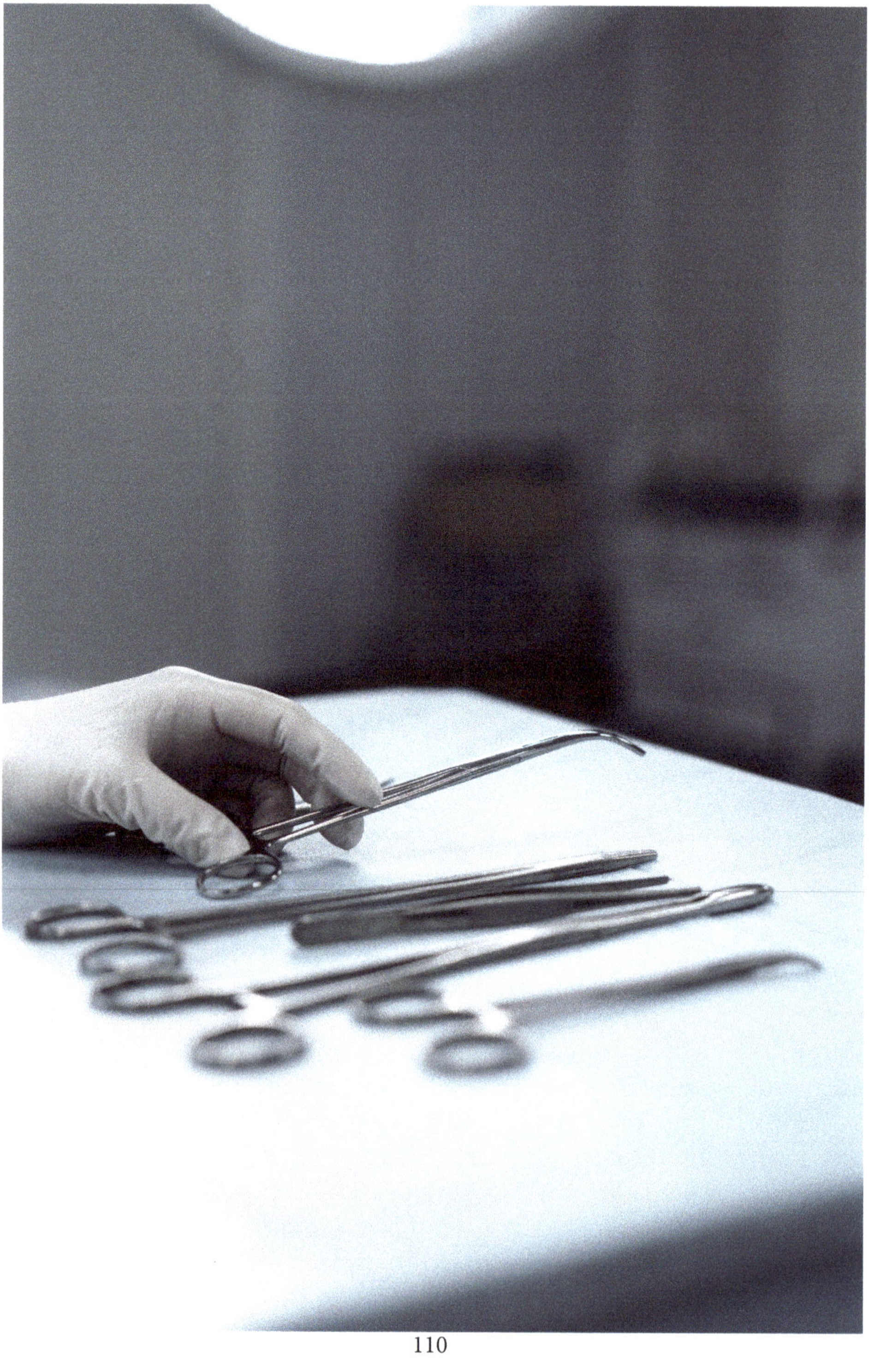

Bibliography

Bibliography:

Astrini, N. (2018). ISO 9001 and performance: a method review. Total Quality Management & Business Excellence, doi: 10.1080/14783363.2018.1524293.

Asadi,J, Easy ISo 13485:2016, Silosa Consulting Group, 2022

Bou-Llusar, J. C., Escrig-Tena, A. B., Roca-Puig, V., & Beltra´n-Martı´n, I. (2005). To what extent do enablers explain results in the EFQM excellence model? International Journal of Quality & Reliability Management, 22(44), 337-353.

Chatzoglou, P., Chatzoudes, D., & Kipraios, N. (2015). The impact of ISO 9000 certification on firms' financial performance. International Journal of Operations and Production Management, 35(1), 145-174. https://doi.org/10.1108/IJOPM-07-2012-0387

Definition of the terms "medical device" and "in vitro diagnostic (IVD) medical device". Global Harmonization Task Force;

Guidelines for regulatory auditing of quality systems of medical device manufacturers – Part 1: general requirements

Global Medical Devices Nomenclature System (GMDN) [website] (https://www. gmdnagency.org, accessed 2 February 2017).

FDA website address: fda.gov

Gigante, N., & Ziantoni, S. (2015). L'edizione 2015 della norma ISO 9001, 2015. Retrieved from:https://www.accredia.it/app/uploads/2015/12/6050_5_L__700_edizione_2015_della_norma_ISO_9001___Arch__Gigante__Dr__Ziantoni.pdf

ISO (2015a). ISO 9001 - Quality management systems – requirements. Geneva: International Organization for Standardization.

Role of standards in the assessment of medical devices. Global Harmonization Task Force; 2008 ISO (2018). ISO 19011 - Guidelines for auditing management systems quality management systems. Geneva: International Organization for Standardization.

ISO (2019). ISO 9000 Family - Quality Management. Retrieved from: https://www.iso.org/home.html.

Wilson, J. P., & Campbell, L. (2018). ISO 9001:2015: the evolution and convergence of quality management and knowledge management for competitive advantage. Total Quality Management and Business Excellence, pp. 1-16. https://doi.org/10.1080/ 14783363.2018.1445965

MDSAP

Vol.1

AUSTRALIA

MDSAP

Vol.2

Brazil

MDSAP

Vol.3

Canada

MDSAP

Vol.4

Japan

MDSAP

Vol.5

USA

www.ingramcontent.com/pod-product-compliance
Lightning Source LLC
LaVergne TN
LVHW051044210926
848486LV00001B/22